The Cowpathy Man

NIPA® GENX ELECTRONIC RESOURCES & SOLUTIONS P. LTD.
New Delhi-110 034

About the Author

Prof. R.S. Chauhan born on September 10th, 1958 in a farmer's family, Dr. Chauhan completed his Bachelors (80.3%) with honours, Master (85.8%) and Doctoral degree (90.0%) in Veterinary Sciences with specialization in veterinary pathology from G.B. Pant University of Agric. & Tech., Pantnagar. He served the country in various capacities including Assistant Professor/ADIO (1983-96), Associate Professor (1996-1999), National Fellow (1999-2004) and Joint Director (CADRAD), IVRI, Izatnagar (2004-2009), Director & Vice-Chancellor (2009), ICAR-IVRI and Campus Director, IBT, Patwadangar (2009-2013). During his tenure as academician and scientist, he has written 112 books including 35 manuals and 1 monograph very popular among the students world over. He contributed 99 chapters in different books and published 235 research and 59 review papers. Besides, he participated in International / National Conferences and presented 178 papers. He is life member of 15 scientific bodies and has been in several executive committee such as Chairman, Panch Gavya Professionals' Club; President, Cow Therapy Society; Secretary-General, Society for Immunology and Immunopathology; Vice President, Indian Society of Veterinary Educators; Zonal Secretary, Indian Association of Veterinary Pathologists and Joint Secretary, Indian Virology Society, President, Dr. J.L. Vegad Foundation, Expert Member Research Group, Committee for Certification of Pathologists, Vice-President, IAVP; Registrar, ICVP, etc.

Based on his contributions and scientific achievements, he has been awarded with several prizes, medals and honours including **Best Young Scientist Award** (1992), IAAVR Award (1996), **National Fellow Award** (1999), Fellow NAVS (2000), Fellow SIIP (2001), K.S. Nair Memorial Award (1999), Vigyan Bharti Award (2000), Dr. C.M. Singh Trust Award (2002), **Dr. Rajendra Prasad Award** (2002), Shri Ramlal Agrawal National Award (2000), Best Paper Award SIIP (2003), Best Paper Award IAVA (2001, 2002, 2003), Best Teacher Award (2004) by GBPUAT, Fellow, IAVP (2006), **Gopal Gaurav** (2007), Bharat Excellence Award (2007), Diplomat, ICVP (2008), Intas-ISVE **Best Veterinary Scientist Award** (2008), **Best Academician Award** (2012), Moropant Pingle **Go Sewa National Award** (2015), Outstanding Scientist Award (2019), Research Excellence Award (2020), Indo Asian-Claude Bourgelat Distinguished Innovative Scientist Award-2020 in Animal Immunopathology and **MAN OF COWPATHY** at GADVASU Ludhiana etc. in recognition of his research and teaching endeavor. He has been inducted in many national and international scientific/advisory committees and boards including Member, WHO/IPCS Committee on Environmental Health Criteria. Dr. Chauhan has advanced scientific field in many new veterinary diseases including Enterotoxaemia, Pyometra, ETEC infection in camels, isolated camel pox virus, developed rapid diagnostic test DIA for the first time in India. He reported role of cell mediated immunity in rotavirus infection in calves. His pioneer work includes immunopathology due to pesticides, heavy metals, mycotoxins and nanoparticles. Dr. Chauhan developed a new method using MTT dye for detection of CMI response. He has scientifically validated Panchgavya and named it as **"Cowpathy"**.

He have been recognized internationally as visiting Professor, University of Wageningen The Netherlands and as Scientific Advisor, The World Health Organization (WHO). Dr. Chauhan guided more than 50 scholars for their Masters and Doctoral research; most of them are placed as Professors, Scientists, Officers in Indian Army, banks and industry in India and abroad. At present he is working as Head Veterinary Pathology, Chairman Scientific Advisory Committee ICMR-NARFBR, Member AWBI and CPCSEA (Govt of India). Prof Chauhan superannuated on 30th June 2024 after distinguish service of 41.5 years and still contributing to the profession through lectures, writings and research and particularly developing literature/ books for the benefit of students and young scientists.

The Cowpathy Man

Prof. Dr. Ramswaroop Singh Chauhan
M.V.Sc., Ph.D. (Path.), FNAVS, FSIIP, FIAVP, PDCR, OCTT, ACPPM, MBA
Ex-Advisor WHO, Ex-Director IVRI, Ex-JDCADRAD, Ex-Director IBT, Ex-National Fellow ICAR
Ex-Chairman, Scientific Advisory Committee, ICMR-NARFBR, Ex-Professor and Head
and Ex-Member, CCSEA (Govt of India)
Member Animal Welfare Board of India (Govt. of India)
Gauma Farm, Inderpur - Pratap Pur Distt. Udham Singh Nagar - 263 148
Uttarakhand, India

NIPA® GENX ELECTRONIC RESOURCES & SOLUTIONS P. LTD.
New Delhi-110 034

NIPA® GENX ELECTRONIC RESOURCES & SOLUTIONS P. LTD.

101,103, Vikas Surya Plaza, CU Block
L.S.C. Market, Pitam Pura, New Delhi-110 034
Ph : +91-11-43860225, Mob.: +91 9717133558, 9540816132
E-mail: newindiapublishingagency@gmail.com
Website: www.nipaersources.com

Print ISBN: 978-93-58879-67-4
ebook ISBN: 978-93-58874-67-9

Composed and Designed by NIPA®.

Contents

1 Introduction 1

2 Understanding Panchgavya: Its Composition, Benefits, and Uses 7

3 Career 15

4 A Great Teacher (GURU) and A Renowned Scientist.........................23

5 The Foundations of Compassion: Early Years....................................29

6 Celebrating Dr. Chauhan's Outstanding Achievement: A Testament to Dedication and Excellence 33

7 A Trailblazer in Veterinary Science: Celebrated Journey of Excellence 41

8 The Legacy of Dr. Chauhan: A Journey of Triumph and Compassion 47

9 Personal Philosophy 51

10 Future Goals of Dr. Chauhan in Expanding the Scope of Veterinary Science 57

11 The Community Engagements of Veterinary Scientist Prof Chauhan in Animal Welfare Activities 61

12 Peer Review: What Others Say About Dr. R.S. Chauhan....................67

13 Cowpathy Publications 81

13 Some Important Moments in The Eyes of Camera.............................91

1

Introduction

Dr Chauhan is a name that resonates with compassion, innovation, and an unyielding commitment to animal welfare. With a career spanning over three decades, he has emerged as a towering figure in the field of veterinary medicine, touching the lives of animals and humans alike through his groundbreaking work and empathetic approach. His exceptional dedication to the health and well-being of cows has earned him the affectionate nickname "Cowpathy Man," a testament to his pioneering efforts in bovine healthcare, conservation of indigenous cows and his relentless advocacy for these gentle creatures.

From a young age, Dr. Chauhan displayed an innate connection with animals. Growing up in Aligarh, U. P. surrounded by rural landscapes and farm animals, he developed a deep appreciation for the role of cows in sustaining communities. This early bond shaped his career path, guiding him toward Veterinary Medicine with a special focus on bovine health. After earning his veterinary degree, he embarked on a journey that would not only establish him as a leading expert in the field but also as a passionate advocate for the humane treatment of animals.

Dr. Chauhan quickly rose to prominence due to his innovative approach to Veterinary care. While he treated animals of all kinds, his work with cows became the cornerstone of his career. Recognizing the cultural, economic, and ecological significance of these animals, he delved into research on common bovine diseases, nutrition, and sustainable farming practices. His dedication led to the development of innovative diagnostic protocols, treatment and preventive measures, improving the health and productivity of cows across the region.

One of Dr. Chauhan 's most notable achievements is his work in promoting "Cowpathy," a holistic approach to bovine care that combines traditional knowledge with modern Veterinary science. This methodology emphasizes natural remedies, ethical practices, and a deep understanding of the unique needs of cows. He is a fierce advocate of popularizing science. Dr Chauhan made complex scientific information accessible and understandable to a general audience, often through plain language, relatable examples, and engaging visuals, while maintaining accuracy and avoiding oversimplification. Through his job portfolio, seminars, publications, and fieldwork, he educated farmers and veterinarians on the benefits of cowpathy, gaining widespread recognition for his efforts. His groundbreaking research in this area has not only enhanced the lives of countless cows but also contributed to the sustainability of dairy farming, benefiting communities dependent on these animals for their livelihood.

The nickname "Cowpathy Man" is more than a title; it represents Dr. Chauhan's unwavering dedication to his mission. Over the years, he has become a symbol of hope for farmers struggling to care for their livestock. His expertise and compassionate approach have transformed the lives of many, from rescuing ailing cattle to devising cost-effective solutions for disease prevention. He has often been seen visiting remote villages, offering free medical camps for cows and training sessions for farmers. His hands-on approach has endeared him to countless communities, earning their respect and gratitude.

Dr. Chauhan's contributions extend far beyond clinical practice. As an academic and researcher, he has authored numerous scientific papers on bovine health, sharing his findings with the global veterinary community. His work has been instrumental in advancing knowledge on issues such as mastitis, foot-and-mouth disease, and reproductive health in cows. In recognition of his contributions, he has been invited to speak at international conferences, where he continues to inspire colleagues and young veterinarians with his insights and dedication. In early 2000, He presented scientific research on cowpathy in Europe and raised awareness, a bold move for defending traditional knowledge in scientific way.

In addition to his professional achievements, Dr. Chauhan is a vocal advocate for animal welfare. He has worked tirelessly to raise awareness about the ethical treatment of cows, campaigning against practices that harm their well-being. His efforts have influenced policy changes and inspired community-led initiatives to create safer, more humane environments for livestock. His advocacy work underscores his belief that the health of animals and the health of communities are deeply interconnected.

Dr. Chauhan's journey has not been without challenges. In a world often driven by profit, his emphasis on ethical practices and holistic care has sometimes faced resistance. However, his unwavering commitment to his principles and his ability to demonstrate the long-term benefits of his approach have silenced skeptics and won over supporters. His resilience in the face of adversity has made him a role model for many in the field.

Among the many accolades Dr. Chauhan has received, a vast number of prestigious Awards, Honours & Accolades, which hold special significance. These recognitions underscore the impact of his work and solidify his legacy as a leader in veterinary medicine. Yet, despite his numerous achievements,

Dr. Chauhan remains humble, attributing his success to the animals he serves and the people who support his mission. As an educator, Dr. Chauhan has inspired countless students to follow in his footsteps. His lectures and workshops are renowned for their depth, accessibility, and passion, often blending cutting-edge research with practical wisdom. Many of his students have gone on to become successful veterinarians, crediting Dr. Chauhan as their greatest mentor and guide. His influence extends far beyond the classroom, as he continues to advise institutions and organizations on the future of veterinary science.

Dr. Chauhan's impact is not limited to the local community; his work has gained international recognition. He has collaborated with global organizations to address pressing challenges in animal health, including the prevention of zoonotic diseases and the promotion of sustainable agricultural practices. His ability to bridge traditional knowledge with modern science has made him a sought-after expert, invited to consult on projects that aim to improve the lives of both animals and people around the world.

In his personal life, Dr. Chauhan is known for his humility and accessibility. Despite his many accomplishments, he remains deeply connected to the communities he serves. He often spends time in rural areas, listening to farmers' concerns and sharing his expertise to address their challenges. His presence is a source of comfort and assurance, reflecting his belief that veterinary medicine is as much about building trust as it is about providing care.

Dr. Chauhan's story is a powerful reminder of the profound impact one individual can have on the world. His work has not only improved the lives of countless animals but also strengthened the bond between humans and the natural world. Through his innovative practices, compassionate care, and relentless advocacy, he has redefined what it means to be a veterinarian, setting a standard of excellence for future generations.

As "Cowpathy Man," Dr. Chauhan has created a legacy that transcends borders and disciplines. His dedication to the well-being of cows and his contributions to veterinary science serve as an enduring inspiration to all who strive to make a difference. Whether in the clinic, the classroom, or the community, Dr. Chauhan continues to lead with compassion, proving that the care of animals is not just a profession but a lifelong calling.

Dr. Chauhan's journey began in the small and obscure village of Aligarh, Uttar Pradesh, where he was born on the 10th of September, 1958. His early life was deeply intertwined with the rhythms of rural India, where his father worked as a farmer, their livelihood solely dependent on agriculture. Farming, while a noble profession, has always been fraught with challenges, from unpredictable weather to fluctuating market prices. Though advancements over time have eased some of these difficulties, new challenges have arisen, leaving many farmers in a precarious situation.

Despite these humble beginnings, Dr. Chauhan exhibited an extraordinary aptitude for academics from a young age. Known for his exceptional intellect and perseverance, he consistently impressed his teachers and peers with his academic brilliance. His early accomplishments laid the foundation for a lifelong pursuit of knowledge, one that would later distinguish him as a leading expert in his field.

Among his many academic achievements, Dr. Chauhan devoted a significant portion of his scholarly work to studying Panchgavya—a traditional Indian system that advocates the use of five products derived from cows: milk, curd, ghee, urine, and dung. This unique field of research complemented his broader studies in veterinary science, reflecting a deep empathy for animals, particularly cows. His compassion for these creatures and his fascination with their role in human lives inspired him to pursue a doctorate in Veterinary Pathology.

In his doctoral research, Dr. Chauhan delved into understanding the intricate health challenges faced by cattle, devising innovative solutions that could improve their well-being. His work was not only academic but also deeply personal, shaped by the values instilled in him during his childhood. Growing up in a religious and culturally rich rural environment, he was profoundly influenced by the stories and teachings of his grandmother and mother. These stories often emphasized the sanctity of cows in Indian culture and their symbolic representation of motherhood, nourishment, and selflessness.

Cows were an integral part of Dr. Chauhan's early life. Even as a young boy, he took great pride in caring for the family's cattle. He would rise early to feed and graze them, ensuring they were healthy and content. His dedication extended to washing them and tending to their ailments with whatever resources were available at the time. These experiences forged a lifelong bond with these gentle creatures and fueled his passion for their welfare.

The cultural and symbolic significance of cows resonated deeply with Dr. Chauhan. In India, cows are revered not only for their economic importance but also for their spiritual and cultural roles. Families in rural areas often live off the produce of their farms, and the products derived from cows—milk, butter, curd, and ghee—are staples of their diet. These products, rich in nutrients, sustain both the young and the old, making cows an essential part of rural sustainability. This reverence for cows contrasts sharply with the symbolism of national animals or birds in other countries. For instance, while the eagle, the national bird of the USA, symbolizes ambition and power, the cow represents a more nurturing and selfless ethos—the pure essence of motherhood.

After earning his doctorate in Veterinary Pathology, Dr. Chauhan embarked on a distinguished career that spanned decades. He held several prestigious positions in research and academia, contributing significantly to the field of animal health and welfare. Yet, amidst these achievements, he never lost sight of his passion for Cowpathy—a holistic approach to healing that emphasizes the therapeutic properties of cow-derived products. His relentless dedication to this field earned him the affectionate nickname "Cowpathy Man," a title he embraced with great pride.

Dr. Chauhan's work in Cowpathy went beyond academic research. He actively advocated for its integration into modern veterinary practices, believing that traditional knowledge and contemporary science could coexist harmoniously. He conducted workshops and seminars, sharing his insights with farmers, veterinarians, and researchers alike. His efforts helped revitalize interest in Panchgavya and its applications in sustainable agriculture and healthcare.

In addition to his professional accomplishments, Dr. Chauhan's life serves as a testament to the values of humility, perseverance, and empathy. He remained deeply connected to his roots, often returning to his village to share his knowledge and experiences with the community that shaped him. He believed that education and awareness were key to empowering rural populations, and he worked tirelessly to bridge the gap between traditional wisdom and modern advancements.

Dr. Chauhan's legacy is one of inspiration and impact. His unwavering commitment to his work and his deep respect for the values of his upbringing have left an indelible mark on the fields of veterinary science and animal welfare. Through his efforts, he has not only advanced scientific understanding but also highlighted the enduring relevance of cultural traditions in addressing contemporary challenges.

Today, Dr. Chauhan's contributions continue to inspire a new generation of researchers and practitioners. His story reminds us that even the most modest beginnings can lead to extraordinary achievements when fueled by passion, dedication, and a desire to make a difference. From a small village in Aligarh to becoming a trailblazer in his field, Dr. Chauhan's journey is a shining example of how one individual's vision and determination can leave a lasting legacy.

2

Understanding Panchgavya Its Composition, Benefits, and Uses

What is Panchgavya?

Panchgavya is an ancient, holistic preparation derived from cows, revered in Hindu culture for its spiritual, agricultural, and medicinal benefits. The term "Panchgavya" comes from Sanskrit, where "Pancha" means five and "Gavya" refers to products obtained from cows. True to its name, Panchgavya consists of five components sourced from cows: three primary ingredients (cow dung, cow urine, and cow milk) and two secondary ingredients (curd and ghee).

These ingredients are mixed in a specific proportion and allowed to ferment, creating a potent concoction. When applied in Ayurveda, this preparation is often referred to as "Cowpathy," a traditional practice believed to harness the natural healing properties of cow- derived substances. In contemporary times, the uses of Panchgavya have extended beyond religious rituals to applications in agriculture, medicine, and even environmental conservation.

The Significance of Panchgavya

In Hindu traditions, cows are venerated as symbols of life and sustenance. As a result, the products derived from them hold immense cultural and spiritual value. Panchgavya is used in various rituals to purify spaces, offer prayers, and seek blessings. Beyond its spiritual aspects, Panchgavya has gained recognition for its practical benefits in multiple fields, particularly in organic farming, animal husbandry, and sustainable living.

Composition and Preparation of Panchgavya

To create Panchgavya the following ingredients are blended

Cow Dung: Rich in organic matter and beneficial microorganisms, cow dung plays a pivotal role in enhancing soil fertility.

Cow Urine: Known for its antimicrobial properties, cow urine acts as a natural disinfectant and growth promoter in agriculture.

Cow Milk: A nutrient-rich component that contributes to the overall potency of Panchgavya.

Curd: Made from cow milk, curd provides probiotics and enhances fermentation.

Ghee: Clarified butter made from cow milk, ghee is a source of healthy fats and antioxidants.

The ingredients are mixed and allowed to ferment for a specified period. The fermentation process enhances the bioavailability of nutrients and introduces beneficial microorganisms, making the mixture highly effective for various applications.

Benefits and Applications of Panchgavya

1.In Agriculture: A Boost for Organic Farming

One of the most significant uses of Panchgavya lies in organic/ natural farming. This natural fertilizer and pesticide eliminates the need for harmful chemicals such as synthetic fertilizers, pesticides, insecticides, and antibiotics. By promoting sustainable farming practices, Panchgavya ensures healthier produce and a safer environment.

Benefits for Plants

Enhanced Growth: Panchgavya stimulates chlorophyll production, resulting in higher photosynthetic activity. This leads to increased plant height, a greater number of branches, and improved fruit yield.

Improved Immunity: The preparation strengthens the plant's immune system, making it more resistant to diseases.

> *Indian cow urine has been described in 'Sushrita Samhita' and 'Ashtanga Sangraha' to be the most effective substance/secretion of animal origin with innumerable therapeutic values. The medicines made from cow urine are used to cure several diseases. Taken in measured quantities, cow urine or gaumutra has a unique place in Ayurveda and is suggested for improving general health. Exploring its antimicrobial activities, it is being used to produce a whole range of Ayurvedic drugs. Cow urine therapy has a long history. It is recognized as 'water of life' or "Amrita" (beverages of immortality), the nector of the God in Vedas, sacred Hindu writing, which is said to be the oldest books in Asia. In India, drinking of cow urine has been practiced for thousands of years. It is an important ingredient of panchgavya, which has been tested by various workers for its immunomodulatory properties and have been reported that it enhances both cellular and humoral immune response*

Cost-Effective Solution: As a natural alternative, Panchgavya reduces dependence on expensive chemical inputs.

Additional claims suggest that the use of Panchgavya in farming can

a) Increase the growth rate of plankton, which is essential for maintaining a healthy aquatic ecosystem.

b) Support poultry farming by improving the health and growth rate of birds.

c) Enhance milk production in dairy cattle when used as a dietary supplement.

2. In Animal Husbandry

Panchgavya is also beneficial for livestock. When included in animal feed, it is said to improve digestion, boost immunity, and enhance overall health. Many farmers have reported increased milk yield in cows after incorporating Panchgavya into their diets. Its natural antimicrobial properties help in preventing infections and promoting better weight gain in poultry and cattle.

3. In Medicine: Ayurveda and Cowpathy

The medicinal benefits of Panchgavya have been recognized in Ayurveda, where it is used as a therapeutic agent. Its components are believed to have detoxifying, rejuvenating, and healing properties. Cow urine, for instance, is known for its ability to cleanse the body and improve digestion. Similarly, cow dung has been used in traditional medicine for wound healing and as an antiseptic. While scientific validation is ongoing, proponents of Cowpathy claim that Panchgavya-based treatments can: Boost immunity and energy levels, treat skin conditions like eczema and psoriasis. Aid in managing chronic illnesses.

Recent researches showed that cow urine enhances the immune status of an individual through activating the macrophages and augmenting their engulfment power as well as bactericidal activity. In poultry, cow urine has been reported to enhance the immunocompetence of birds and provide better protection along with vaccination and increases egg production and egg quality. In-vivo cow urine treatment to developing chicks marginally upregulated the lymphocyte proliferation activity. The cow urine (Kamdhenu ark' / cow urine distillate) is a potent and safe immunomodulator, which increases both humoral and cell mediated immunity in mice. It was observed that cow urine enhances both T and B cell proliferation and also increases the titre level of IgG, IgA and IgM antibodies. It increases the secretion of interleukin-1 and 2 also. . The level of both IL-1 and 2 in mice got increased by 30.9 and 11.0%, respectively and in rats these levels were increased significantly by 14.75 and 33.6%, respectively.

4. Environmental Benefits

Panchgavya, a time-honored system rooted in the principles of Ayurveda, has emerged as a premium offering in the healthcare and cosmetics sectors. This holistic approach, derived from five key ingredients sourced from cows—milk, curd, ghee, urine, and dung—has demonstrated immense potential in promoting overall health and well-being. With its natural, eco-friendly, and scientifically supported benefits, Panchgavya-based products are gaining global recognition for their effectiveness in skincare, healthcare, and even chronic disease management.

Premium Products in Healthcare and Cosmetics

The Panchgavya range includes a luxurious line of healthcare and cosmetic products that cater to modern needs while preserving traditional wisdom. In the cosmetics division,

> *Cow urine possesses anti-cancer properties. Studies highlight the role of cow urine in curing cancers and that cow urine enhances the efficacy and potency of anti-cancer drugs. Scientists have proved that the pesticides even at very low doses cause apoptosis (cell suicide) in lymphocytes of blood and tissues through fragmentation of DNA. Distilled cow urine protects DNA and repairs it rapidly as observed after damage due to pesticides. It protects chromosomal aberrations by mitocycin in human leucocyte. Cow urine helps the lymphocytes to survive and not to commit suicide (apoptosis). It has been reported the prevention of pathogenic effect of free radicals through cow urine therapy. These radicals cause damage to various tissues and attack enzymes, fat and proteins disrupting normal cell activities or cell membranes, producing a chain reaction of destruction leading to the ageing process of a person. By regular use of cow urine one can get the charm of a youth as it prevents the free radical formation. In-vitro study on Animal cell culture showed the anti-cancerous effect of Taxus baccata and CUD alone and in combination. Out of various combinations, aqueous extract of Taxus baccata and CUD was found most promising anti-cancerous preparation in in-vivo studies.*

there is a vast selection of astringents, creams, lotions, moisturizers, facial kits, and soaps, all formulated using natural ingredients. These products not only rejuvenate and nourish the skin but also provide long-lasting hydration, helping users achieve a youthful glow. The facial treatment kits, for instance, are enriched with herbal and Panchgavya extracts, making them suitable for all skin types. They are designed to detoxify the skin, remove impurities, and promote natural radiance. Haircare products such as shampoos, conditioners, and styling gels are infused with natural oils and cow-derived ingredients, ensuring nourishment, reduced hair fall, and enhanced hair texture.

What sets these products apart is their ability to leave a positive impression on the user's mind and body. The subtle yet invigorating fragrances of herbs and natural oils offer a calming sensory experience, while the therapeutic properties provide tangible results. This unique combination of physical and mental benefits makes Panchgavya cosmetics a preferred choice for individuals seeking holistic skincare solutions.

Healthcare Benefits of Panchgavya Therapy (Cowpathy)

Panchgavya therapy has also made significant strides in the field of healthcare. The oral consumption of products derived from Panchgavya introduces live microorganisms into the body, which act as natural probiotics. These microorganisms play a pivotal role in boosting the immune system by stimulating the production of antibodies. This process resembles the mechanism of vaccines, as it trains the immune system to recognize and combat harmful pathogens.

Cowpathy include cow cuddling, Backrub, Gau grass, Gauseva, and living with cow besides, Panchgavya products milk, curd, ghee, urine and dung. Since the cowpathy practice is to increase the aura of an individual, it prevents from many ailments and infections. Indigenous cow is considered as sacred in Indian literature because its products like urine, dung, milk, curd and butter has many wonderful medicinal properties

Gau grass

The therapeutic applications of Panchgavya extend to a wide range of diseases and ailments. It is particularly recommended for managing conditions such as asthma, flu, allergies, and cardiovascular diseases. Patients suffering from chronic inflammatory conditions like rheumatoid arthritis have reported significant relief after undergoing Panchgavya therapy. Skin disorders such as leucoderma and wounds that are difficult to heal have also shown remarkable improvement with the use of Panchgavya-based treatments. The antibacterial and antifungal properties of cow urine and dung extracts help cleanse the affected areas, promoting faster recovery and reducing the risk of infections.

In the realm of gastrointestinal health, Panchgavya therapy has proven effective in addressing dietary issues, improving digestion, and alleviating symptoms of gastrointestinal diseases. By restoring the natural balance of gut flora, these products enhance nutrient absorption and support overall digestive health.

Back rub

Potential Against Chronic Diseases

One of the most intriguing aspects of Panchgavya therapy is its potential to combat chronic and life-threatening conditions. Preliminary studies and anecdotal evidence suggest that Panchgavya may play a role in managing diseases like cancer, AIDS, and diabetes.

In the case of cancer, the bioactive compounds present in Panchgavya have shown antioxidant and anti-inflammatory properties. These compounds help neutralize free radicals, reduce oxidative stress, and inhibit the growth of cancerous cells. Similarly, in AIDS, Panchgavya therapy supports the immune system, improving the body's ability to fight infections and maintain overall health.

For individuals with diabetes, Panchgavya products help regulate blood sugar levels by enhancing insulin sensitivity and reducing glucose absorption. Additionally, the therapy's detoxifying effects aid in cleansing the body of harmful toxins, which is particularly beneficial for diabetic patients prone to complications.

Holistic Wellness and Mental Health

Beyond its physical health benefits, Panchgavya therapy also contributes to mental and emotional well-being. The natural aroma of cow-derived ingredients, combined with the soothing effects of herbalpreparations provide holistic heath including mental health.

Beyond its agricultural and medicinal uses, Panchgavya plays a role in promoting environmental sustainability. Its use in farming reduces chemical runoff into water bodies, thereby preventing pollution. Moreover, the practice of utilizing cow by-products aligns with principles of zero waste, ensuring that every part of the cow's output is put to good use.

The Science Behind Panchgavya

The efficacy of Panchgavya can be attributed to its rich microbial diversity. The fermentation process introduces beneficial bacteria and enzymes, which contribute to its effectiveness in enhancing plant growth, improving soil fertility, and combating pathogens. These microbes also play a role in breaking down organic matter, releasing nutrients that plants can readily absorb.

Research has shown that Panchgavya-treated plants exhibit higher levels of chlorophyll, improved root development, and better resistance to pests and diseases. Similarly, its antimicrobial properties make it an effective natural disinfectant in livestock farming.

Challenges and Future Prospects

Despite its numerous benefits, the use of Panchgavya faces certain challenges. For instance, its preparation requires careful attention to detail, as improper fermentation can lead to ineffective or harmful results. Moreover, the adoption of Panchgavya in mainstream agriculture and medicine is limited due to a lack of awareness and scientific validation.

To address these issues, ongoing research is essential. Efforts should also be made to promote the use of Panchgavya through training programs, government initiatives, and collaboration with agricultural and medical communities.

Conclusion

Panchgavya is a testament to the wisdom of traditional practices, offering sustainable solutions for modern challenges. Its versatility and effectiveness make it a valuable asset in agriculture, animal husbandry, medicine, and environmental conservation. By embracing and refining this ancient preparation, we can pave the way for a healthier, more sustainable future.

3

Career

The professional journey began in 1983, when Dr. Chauhan joined as an Assistant Professor / Assistant Disease Investigation Officer (ADIO). This role marked the start of a career dedicated to education, research, and the betterment of scientific inquiry. Over the next 13 years, he immersed himself in the dual responsibilities of teaching and organizational leadership. These formative years were characterized by a commitment to mentoring students, developing academic programs, and fostering an environment of learning and discovery. It was a period of establishing strong foundations in both pedagogy and administration, shaping his approach to addressing challenges with strategic foresight and innovation.

In 1996, he transitioned to a role focused solely on academic excellence as an Associate Professor, continuing until 1999. This phase allowed him to refine his teaching methodologies, expand the scope of his research, and contribute meaningfully to academic advancements. It was a period of growth, both personally and professionally, as he worked closely with students and peers to push the boundaries of knowledge in his field.

A pivotal moment in his career came in 1999, when he was appointed as a National Fellow. This prestigious designation offered a platform to undertake pioneering research and address some of the most pressing challenges in his area of expertise. Over the course of five years, from 1999 to 2004, he delved deeply into advanced research projects, focusing on innovation and impactful outcomes. This period was marked by the development of solutions that not only advanced scientific understanding but also had practical applications, benefitting both the academic community and society at large.

In 2004, he took on a leadership role as the Joint Director at the Centre for Animal Disease Research and Diagnosis (CADRAD), Indian Veterinary Research Institute (IVRI), Izatnagar. His tenure at IVRI, spanning from 2004 to 2009, was a dynamic phase of his career. As Joint Director, he was tasked with overseeing critical research initiatives, guiding teams of talented scientists, and implementing strategic plans to advance the institute's mission. This period was defined by significant milestones, including the successful execution of projects that enhanced disease diagnostics and reproductive technologies in the veterinary sciences. His leadership not only contributed to institutional growth but also to the broader scientific community's understanding of animal health and productivity.

In 2009, he ascended to the role of Director and Vice Chancellor at the Indian Council of Agricultural Research (ICAR) – IVRI. His tenure as Director and Vice Chancellor was marked by a focus on fostering interdisciplinary collaboration, strengthening research infrastructure, and building partnerships with national and international organizations. These efforts not only elevated the reputation of ICAR-IVRI but also positioned it as a leader in agricultural and veterinary research.

Thereafter, he assumed the position of Campus Director for Biotechnology at Padwanagar. These roles represented the pinnacle of his career, as they required a unique blend of vision, leadership, and expertise. Over the next four years, from 2009 to 2013, he was responsible for steering these esteemed institution toward new heights of academic and scientific achievement.

As Campus Director for Biotechnology, he played a key role in integrating cutting-edge technologies into research and education. He was particularly passionate about nurturing the next generation of scientists and leaders. By fostering a culture of curiosity and innovation, he ensured that students and researchers were equipped with the skills and knowledge necessary to tackle the complex challenges of the future. This dual responsibility allowed him to contribute holistically to the development of both institutions and their stakeholders.

Throughout his career, he had consistently sought to bridge the gap between academic research and practical applications. Whether it was through mentoring students, leading research projects, or managing institutions, his focus had always been on making meaningful contributions to society. He had been fortunate to work alongside some of the brightest minds in the field, and their collaboration had been instrumental in achieving success at every stage of his journey.

Each role, which has have undertaken has been a stepping stone, contributing to a cumulative reservoir of knowledge, experience, and insight. From the early days as an Assistant Professor and ADIO to leading premier institutions as Director and Vice Chancellor, his professional journey has been one of continuous learning, growth, and impact. These decades of service have been as fulfilling as well as challenging, offering countless opportunities to make a difference in education, research, and institutional growth.

Looking back, he does take a certain pride of the milestones achieved and the legacy of excellence that his teams and he had built. The journey has not only been a testament to hard work and perseverance but also to the power of collaboration and shared vision. It is his sincere hope that the contributions made over the years will continue to inspire and guide future generations of educators, researchers, and leaders.

Dr. Chauhan, a distinguished Academic and Scientist, has established a remarkable legacy through his extensive contributions to education, research, and the dissemination of scientific knowledge. His unparalleled dedication to his field is reflected in his prolific output of written works. Over the course of his illustrious career, Dr. Chauhan has authored an astounding 120 books, covering a wide range of topics relevant to his discipline. These books, known for their comprehensive and insightful content, have earned him both national and international recognition among students, researchers, and professionals alike. In addition to these publications, he has also written 35 instructional manuals that have served as essential guides for practitioners and students, along with one highly acclaimed monograph that has set a benchmark in the field.

Beyond authoring books, Dr. Chauhan has contributed 107 chapters to various edited volumes, showcasing his expertise in diverse subfields. His involvement in collaborative projects has allowed him to enrich the academic community and bring forward innovative perspectives. Moreover, he has published an impressive 241 original research papers and 65 critical review articles in peer-reviewed journals. These publications not only demonstrate his commitment to advancing scientific knowledge but also underscore his influence in shaping the direction of research in his area of specialization.

A passionate advocate for the communication of science, Dr. Chauhan has actively participated in numerous national and international conferences. During these prestigious gatherings, he has delivered presentations on 182 research papers, sharing his findings and engaging with fellow scientists from across the globe. His ability to convey complex scientific concepts in an accessible manner has made his presentations both impactful and memorable.

Recognizing the importance of bridging the gap between scientific research and the general public, Dr. Chauhan has also focused on simplifying the language used in scientific literature. This effort has resulted in the publication of 328 semi-scientific articles, which have been instrumental in making science more approachable and relatable for non-specialist audiences. His articles, written in a clear and engaging style, have inspired many individuals to develop an interest in science and its applications. Furthermore, he has created 19 educational pamphlets that address practical aspects of science, offering valuable information to farmers, students, and community members.

Dr. Chauhan's contributions extend far beyond writing and publishing. He has dedicated significant efforts to educating and training a wide array of audiences, including scientists, veterinary officers, para-veterinary staff, farmers, butchers, and fish breeders. His systematic approach to teaching and his ability to tailor content to meet the needs of diverse groups have made him an exceptional educator. Over the years, he has conducted more than 19 specialized training programs, covering topics that range from advanced scientific techniques to practical, hands-on skills. These programs have played a vital role in enhancing the knowledge and competencies of participants, thereby contributing to the overall development of their professions.

In addition to his work as an educator, Dr. Chauhan has been a prominent voice in the media, leveraging platforms like All India Radio and Doordarshan to discuss important topics related to his field. His radio and television appearances have enabled him to reach a broader audience, furthering his mission to popularize science and raise awareness about its relevance in everyday life. Through these engagements, he has addressed critical issues and provided practical solutions, earning the admiration and respect of listeners and viewers across the country.

Dr. Chauhan's approach to science is characterized by a deep understanding of its societal implications. He firmly believes in the transformative power of knowledge and has consistently worked to ensure that scientific advancements benefit not only researchers and professionals but also the wider community. His ability to connect with people from various walks of life has been instrumental in fostering a greater appreciation for science and its potential to improve lives.

A testament to his exceptional communication skills and commitment to excellence, Dr. Chauhan's work has earned him widespread acclaim. His books and articles are not only widely read but also frequently cited by other scholars, reflecting their significance in the academic world. His training programs and public outreach initiatives have left a lasting impact, empowering individuals and communities to make informed decisions based on scientific principles.

As a speaker, Dr. Chauhan has been invited to address audiences at prestigious conferences and events, where his presentations have been lauded for their clarity, depth, and relevance. His ability to engage with audiences of varying levels of expertise has set him apart as a thought leader and an inspiring figure in his field. Dr. Chauhan's dedication to simplifying science and making it accessible to the common people is one of the hallmarks of his career. He has consistently emphasized the importance of communication in science, recognizing that the true value of knowledge lies in its ability to create positive change. Through his semi-scientific articles and public discussions, he has demystified complex concepts, making them relatable and actionable for a wide audience.

In conclusion, Dr. Chauhan's contributions

Dr. Chauhan is a highly accomplished professional with an extensive career dedicated to advancing veterinary science, education, and innovative practices in animal health. His leadership roles in numerous esteemed organizations and his contributions to research and education reflect his unwavering commitment to the field. Over the years, Dr. Chauhan has earned Life Memberships and has been an integral part of several key executive committees, making a lasting impact on the veterinary and scientific community.

Achievements & Awards

At present, Dr. Chauhan serves as the Chairman of the Panchgavya Professionals Club, where he actively promotes the benefits of Panchgavya (a traditional Indian agricultural and medicinal preparation derived from cow products, and its applications in sustainable agriculture and holistic healthcare. As the President of the Cow Therapy Society, he champions the therapeutic potential of bovine- based remedies, focusing on integrating traditional practices with modern science to create effective treatments for various ailments.

Dr. Chauhan's contributions extend beyond these specialized fields. He serves as the Secretary General of the Society of Immunology & Immunopathology, where he plays a pivotal role in advancing research and innovation in immunological studies and their applications in veterinary and medical sciences. Additionally, his position as Vice- President of the Indian Society of Veterinary Educators underscores his commitment to fostering excellence in veterinary teaching and training, ensuring that future generations of veterinarians are well-equipped to tackle emerging challenges in animal health.

As the Zonal Secretary of the Indian Association of Veterinary Pathologists, Dr. Chauhan facilitates collaboration among veterinary pathologists across regions, promoting knowledge sharing and advancements in diagnostic pathology. In his capacity as Joint Secretary of the Indian Virology Society, he contributes to the study of viruses affecting animals, ensuring the development of effective prevention and control measures for viral diseases.

Dr. Chauhan also serves as the President of the Dr. J. L. Vegad Foundation, an organization dedicated to commemorating the legacy of Dr. J. L. Vegad, a pioneer in veterinary pathology. Under his leadership, the foundation has undertaken initiatives to support research, provide scholarships, and recognize outstanding contributions in the field of veterinary pathology. His expertise is further recognized through his role as an Expert Member of the Research Group and Committee for the Certification of Pathologists, where he provides valuable insights into the evaluation and accreditation of veterinary pathologists. Dr. Chauhan is also a prominent figure in the Indian Association of Veterinary Pathologists (IAVP) and the Indian College of Veterinary Pathologists (ICVP), where he holds dual roles as Vice-President and Registrar. These positions highlight his dedication to upholding the highest standards of veterinary pathology and promoting professional development within the field.

Dr. Chauhan's career is marked by his ability to seamlessly integrate leadership, research, and advocacy. He has been instrumental in promoting interdisciplinary collaboration, bridging traditional and modern scientific approaches, and addressing critical issues in veterinary science. His work in immunology, virology, pathology, and veterinary education has earned him widespread recognition and respect from peers and institutions alike.

Beyond his formal roles, Dr. Chauhan is an advocate for sustainable practices and ethical approaches in veterinary care. His initiatives in Panchgavya and cow therapy reflect his deep- rooted belief in the importance of aligning scientific advancements with ecological harmony and traditional wisdom. By blending these perspectives, he has set a benchmark for innovation and sustainability in veterinary practices.

Dr. Chauhan is not only a leader but also a mentor to aspiring veterinarians and researchers. He actively engages in capacity- building programs, workshops, and conferences, sharing his knowledge and inspiring the next generation of professionals. His contributions to the development of veterinary science curricula and his involvement in policy formulation have further strengthened the foundation of veterinary education in India.

In addition to his professional achievements, Dr. Chauhan is a prolific author and speaker. He has delivered numerous lectures and presentations at national and international forums, addressing topics ranging from immunopathology and virology to traditional veterinary practices. His publications have significantly contributed to the scientific literature, offering valuable insights into emerging trends and challenges in veterinary science.

Dr. Chauhan's multifaceted career is a testament to his dedication, vision, and leadership. His ability to excel across diverse domains and his relentless pursuit of excellence have made him a role model for professionals in the veterinary and scientific communities. Through his work, he continues to inspire innovation, foster collaboration, and pave the way for a brighter future in animal health and welfare.

An academic, researcher, writer, and educator, he has left an indelible mark on his field. His impressive body of work, encompassing books, articles, training programs, and public engagements, reflects his unwavering commitment to advancing knowledge and improving lives. As a scientist who has seamlessly combined rigorous research with effective communication, Dr. Chauhan serves as an inspiring example of how science can be a force for good in society. His legacy continues to inspire students, professionals, and the general public, ensuring that his impact will be felt for generations to come.

4

A Great Teacher (GURU) and A Renowned Scientist

Soon after completing masters in Veterinary Pathology, Dr Chauhan joined as Teaching Associate in the department of Microbiology and Public Health in Pantnagar and taught BVSc &AH students systemic Bacteriology, meat hygiene and milk hygiene. However, this period was very short of only few months as he joined as Assistant Professor/ Assistant Disease Investigation Officer in HAU Hisar (1983-1996) and was entrusted disease investigation besides teaching of UG and PG students. He taught hygiene to BHSc students, and Epidemiology, Zoonosis, Livestock and Poultry Disease Investigation courses to masters and doctoral students. He also guided several masters and doctoral thesis research on pesticide toxicity, reoviral infection, and protozoan parasite infections. He has written a book on "Text book of Veterinary Clinical and Laboratory Diagnosis which was released by Hon'ble Central Education Minister Km Shailja ji. Students were fond of Dr Chauhan because of his teaching skills, easy to learn process and friendly behaviour; among them are Dr Naresh Jindal, Dr NK Mahajan, Dr SK Khurana, Dr Rajsh Khurana, Dr RK Vaid, Dr Devendra Bist etc.

In Feb1996 he joined at Pantnagar as Associate Professor and was given responsibility of Veterinary Pathology teaching to BVSc&AH, MVSc and PhD students. The teaching skills of Dr Chauhan were appreciable and were very useful to the students. He prepared power point presentation of each and every topic with photographs, line drawings, video and graphs/ bar diagrams etc to make it more understandable to the students. Particularly Dr Chauhan's teaching include the Indian perspectives including animal diseases which showed the right things present in the field. Many disease conditions are different in India in comparison to other countries; some important diseases present in India are mentioned in the books of foreign writers and similarly many diseases occurs in other countries are not having any presence in India. So it is very important to impart education to students in right perspective which they will face in the field. Dr Chauhan always discusses the epidemiology part of the disease whether the disease in present in India or not; if present then what is the magnitude of disease. Such information is not available in literature, it comes by virtue of experience.

Students become fond of Dr Chauhan due to is teaching methodology, impartial behaviour and assessment. During this period Dr Chauhan also written several books on Pathology, Immunology, Immunopathology, Zoonosis, and environmental pollution which are very popular among students and veterinarians. Dr Chauhan also initiated several research projects particularly on environment pollution and health, and on cowpthy. The great contributions of Dr Chauhan were recgnised at national level and he was promoted to the post of NATIONAL FELLOW by Indian Council of Agriculture Research. Similarly at international level he was recognised initially as member and then Advisor to World Health Organisation (WHO). He along with other colleagues made a reference book for WHO named as Envionmental Health Criteria"-236 on Autoimmunity due to chemicals; which was distributed world wide and well accepted and read by the people.

While working on adverse effects of pesticides and heavy metals, the ICAR and govt of India were approached for reduction and no use of these agrochemicals but failed due to their ever incrreasing hope for enhanced grain production to feed the people of India. In that scale the authorities refused to shift on natural farming; however, we developed models for natural farming using cow products. In such case it is thought that we should develop such natural products which can wash the harmful effects of agrochemicals. Panchgavya (Cowpathy) was cosidered is one of them besides the herbs. In our ancient literature it was mntioned that indigenous cow urine can detox body as it has been used in Ayurveda since time immemorial to washout the toxic effects of herbs and the process is known as bhavna. Keeping this in mind Indigenous cow urine was used after processing and initial experiments were conductucted in rats, mice,rabbits and/or chickens which was found to be an excellent immunomodulator. Further studies also confirmed this and also reported indigenous cow urine as anti cancer, bioenhancer, immunomodulator, ant infection, and aphrodisiac.

Dr Chauhan got many przes, awards and recognitions as best teacher award, excellent researcher award, Padm sri samman etc. for his best teaching and research abilities. He was also nominated for management committee of AIIMS New Delhi as Member. In the year 2004 he was selected as Joint Director CADRAD IVRI where he worked till 2009. During this period he was entrusted many responsibilities for better administration and fuctioning of the institute which he completed with great honour. Here also Dr Chauhan was associated with teaching of epidemiology, disease investigation and trainees besides PG students and also guided several students for their mastres and doctoral research. He was famous for his teaching skills among the students and others.He reshaped the CADRAD , shifted in new building make it a 24x7 service for the welfare of animals and their owners. The final diagnosis time was reduced to 72 hrs with initial reporting within 24 hrs of receiving samples.

Progress Report of IBT, Patwadangar
(Sept 2009- June 2013)

1. Training Programmes organized : 31
2. Young Scientists Trained : 281
3. Research Projects undertaken : 06
4. Project Training Work : 225
5. Publications : 59
    a. Books – 09
    b. Manuals – 18
    c. Booklets – 08
    d. Research papers – 14
    e. Folders – 07
    f. Technical Bulletins – 03
6. Seminar/Conferences/Symposia organized :09
7. International cooperation/visiting scientists:

i. Dr. Nicodemus Useh, Nigeria February 3-27, 2011.
ii. Dr. C. Putcha, U.S.A., June 2012.

For few months, he was also appointed as Director IVRI which he served till April 2009. Dr Chauhan as Director IVRI allowed to purchase new vehicles like ambulance for hospital, bus for students and extension activities and staff car for office of Director wich was a major breakthrough as the vehicles were not purchased since last 10 years. Besides teaching, researc and administration work set on time and run smoothly. Meetings were sheduled as per time allotted with no extra gup-sup, adhere to agenda items and quick decisions were tacken which made IVRI famous for its fast actions and functions for the benefit of animal owners/ farmers.

After the distinguished 5 years service in IVRI, Dr Chauhan joined Institute of Biotechnology at Patwadangar Nainital where he served again for 4-5 years. He made the institute vibrant with scientific research, training and extension activities; trainees were coming from different forign countries like USA, UK, Nigeria, Nepal etc. besides from various universities from India. Dr Chauhan also organised several scientific conferences at IBT and wrote several books. Many students of Biotechnology came over there for their masters and / or doctoral research. Cowpathy research was also actively done in Bioprospecting laboratory in the campus; Chief minister and other ministers usually visit the institute and always appreciated the work of Dr Chauhan.

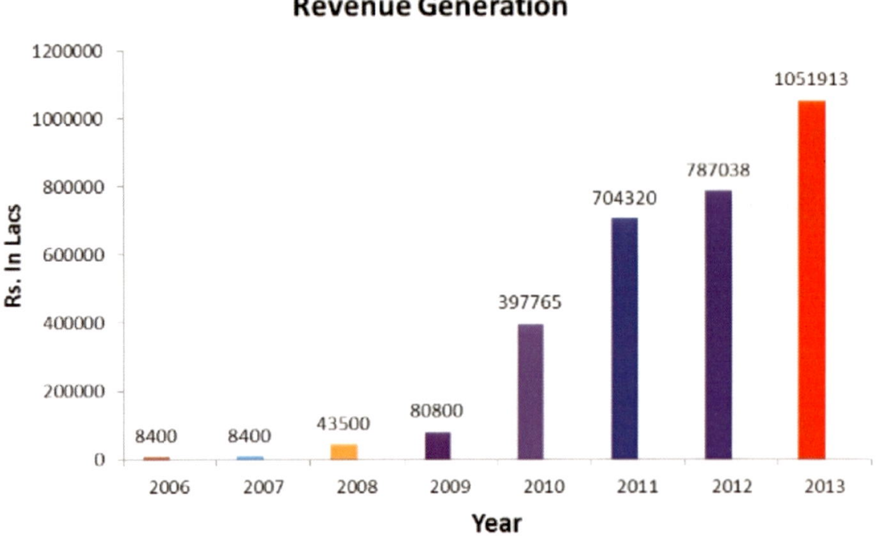

In 2013 end Dr Chauhan joined back in the department of Pathology as Professor and Head and remained continued till his superannuation in June 2024. During this period Dr Chauhan continued his teaching and research work, writing of books and extra mural research projects from industry, UAS, govt of Karnataka and ICAR worth crores of rupees; of these 2 projects were on Cowpathy while third one was on Effects of Bt cotton seeds on lactating buffaloes. It has been found that the Bt cotton is having negative health effects on buffaloes. Cowpathy research yielded new information that leads to few patents also. Research on Cowpathy using Badri cow urine from hills and plain areas were showing positive health effects on hematological , biochemical, liver function, kidney functions, and immunological parameters leading to a good immunomodulation, which can be exploited for the development of an immunomodulatory preparation. In fact the further research on this aspect included the conversion of Badri cow urine into powder form, preparation of various combinations with herbs and study of these combinations for their biological activity in experimental animals like rats. The results were highly encouraging and will be utilized for preparing final report of the project, filing of patents and publication of research papers in journals.

Training under Dr. Chauhan has been one of the highlights of my career, and I take great pride in it. For all that he has taught me and for the immense impact he has had on my life, I will be forever grateful. He is an inspiration, a guiding force, and, of course, a true Guru!

गुरूर ब्रह्मा, गुरूर विष्णु, गुरूर देवो महेश्वरः

Dr. Bhanu Pratap Singh
Senior Director, Nonclinical Safety and Pathobiology
Gilead Sciences, Inc. CA, USA

During this period Dr Chauhan guided several students for their masters and doctoral degree who are working well in the universities, government and industry as reputed Pathologists.

5

The Foundations of Compassion Early Years

Dr. Chauhan, a celebrated veterinary doctor whose groundbreaking work in rural animal healthcare earned him international accolades, began his journey in a modest setting. Born in 1958 in a small village near Aligarh, Uttar Pradesh, Amit grew up surrounded by fields, animals, and the rustic charm of rural India. This environment would unknowingly shape his path toward becoming one of the most sought-after veterinary doctors in the country.

A Humble Beginning

The Chauhan family owned a small plot of farmland and a handful of livestock, which formed their primary source of livelihood. His father, was a hardworking farmer who toiled from sunrise to sunset in the fields, while his mother, Prakash Vati Devi, managed the household and cared for their cows and buffaloes. Despite their modest means, the family placed immense value on education. Ramswaroop, an Intermediate passout in his youth which was quite an achievement few in the village had accomplished—dreamed of giving his children a brighter future.

Dr. Chauhan's early years were spent observing the rhythm of village life. The crowing of roosters, the mooing of cows, and the occasional bleat of goats were the background score of his childhood. While other children enjoyed chasing buffaloes or playing in the fields, Dr. Chauhan was different. His deep empathy for animals was evident even as a child, as he often helped his mother care for sick cows or newborn calves. It was during these moments that the seeds of his future profession were sown.

Early Education: The Village Primary School

At the age of five, Dr. Chauhan was enrolled in the government primary school in the village, a small building with cracked walls, a tin roof, and a single blackboard shared by all students. The school had only two teachers who managed Classes 1 to 5. Resources were scarce—students sat on jute mats, and textbooks were often passed down from older siblings. Yet, the school was a beacon of hope for many families who aspired for their children to break free from the cycle of poverty.

Dr. Chauhan was a diligent and curious student from the start. His teacher, Master Harpal Singh, quickly noticed his keen intellect and love for learning. Despite the lack of resources, he excelled in his studies, often helping his classmates understand difficult concepts. His favorite subject was environmental studies, particularly lessons involving plants and animals. When the teacher spoke about the life cycle of a butterfly or the anatomy of a cow, Dr. Chauhan listened with rapt attention, his imagination painting vivid pictures.

The Spark of Compassion

One summer evening, when he was seven, a stray dog wandered into the village with a severe wound on its leg. While most villagers shooed the dog away, Dr. Chauhan was determined to help. He brought the dog home, cleaned its wound using turmeric and neem paste, and fed it scraps of roti. Over the next few days, the dog began to recover, and Dr. Chauhan proudly declared it his "first patient." This act of kindness earned him admiration in the village and solidified his connection to animals. Ramswaroop began spending more time observing the village animals. He noticed how a mother hen protected her chicks, how cows reacted when they were unwell, and how goats playfully interacted with one another. These observations deepened his understanding of animal behavior and strengthened his desire to care for them.

Intermediate School: A Wider World

When he completed Class 5, his parents faced a tough decision. The village school did not offer education beyond the primary level, and the nearest intermediate school was 4 kilometers away in a neighboring town. Sending Ramswaroop to this school meant financial strain and a logistical challenge, as he would have to walk to school every day. However, his parents were unwavering in their resolve to support Dr. Chauhan's education. At the age of thirteen, Dr. Chauhan enrolled at Gandhi Inter College Chandaus, a government managed school known for its disciplined environment and relatively better infrastructure. The first day of school was both exciting and intimidating for him. Unlike his village school, Gandhi Inter College Chandaus had proper classrooms, desks, and even a small library. The transition from his small village school to a more structured environment was not easy, but Dr. Chauhan quickly adapted.

Academic Excellence and Early Mentorship

At Gandhi Inter College Chandaus, his talent truly began to shine. His science teacher, Mr. Harpal Singh, became his mentor and a significant influence in his life. Recognizing his keen interest in biology, he encouraged him to participate in science fairs and take an active role in the school's nature club. Under her guidance, Ramswaroop developed the skill of writing notes, managing time of study at home, reading and memrise the things etc.

> *It is the true incident of around 1975 when I was studying in class 12. On one day during assembly and prayer time, the principal of our college Shri SC Gupta ji told the audience that this student (Ramswaroop Singh Chauhan) will gloryfy our college name. This incident, I always remember whenever I got any award/ recognition and I always dedicate to my teachers who laid foundation of my career.*

Mr. Singh also introduced Ramswaroop to the concept of animal husbandry and veterinary medicine. During a classroom discussion on careers, he mentioned, "Veterinary doctors are as important as human doctors. They not only treat animals but also ensure the well-being of families who depend on these animals for their livelihood." This was a pivotal moment for him as it gave him a clear vision of his future.

Challenges and Determination

The journey to intermediate school was arduous, especially during the monsoon season when the unpaved roads turned into muddy swamps. There were days when Dr. Chauhan arrived at school soaked to the skin, yet his resolve never wavered. At home, he often studied by the dim light of a kerosene lamp, as electricity was not available in the village. Financial difficulties also loomed large. His parents occasionally had to sell a portion of the family's harvest to pay for Dr. Chauhan's school fees and books. Despite these hardships, his parents never let him feel the weight of their sacrifices. Their unwavering support fueled his determination to excel.

Bridging Theory and Practice

During holidays and off time , Dr. Chauhan assisted the parents in raising animals at home where he also learned how to feed, milk, and graze the animals. He also learned how to recognize symptoms of common ailments like mastitis and foot-and-mouth disease in cows and buffaloes. One memorable incident occurred when his father treated a buffalo suffering from tympany. Watching this treatment of the animal with precision left a profound impression on him. "This is what I want to do," he told himself that day.

A Vision for the Future

By the time Dr. Chauhan completed his intermediate education, he was among the top-ranking student in the state. His academic achievements earned him a scholarship from ICAR to pursue higher education in Pantnagar, a significant step toward realizing his dream. As he prepared to leave his village, Ramswaroop reflected on the journey so far. He owed his success to the unwavering support of his parents, the mentorship of his teachers, and the lessons learned from the animals he had cared for. As he bid farewell to his village, he carried with him not just textbooks and dreams but also a deep sense of responsibility. He was determined to return one day and use his education to improve the lives of the people and animals in his village.

Lessons from the Journey

Dr. Chauhan's early years were shaped by the simplicity of village life, the challenges of rural education, and the boundless support of his family and mentors. These experiences instilled in him a sense of resilience, empathy, and purpose. They also reinforced his belief that education and compassion could transform lives. Today, Dr. Chauhan often reflects on those formative years with gratitude. His journey from a curious village boy to an awarded veterinary doctor is a testament to the power of determination, community support, and the unbreakable bond between humans and animals.

6

Celebrating Dr. Chauhan's Outstanding Achievement: A Testament to Dedication and Excellence

Dr. Chauhan is an exceptional individual who had achieved a significant milestone in his academic and professional journey. Completing a Bachelor of Science (Honours) with an impressive 80.3%, Dr. Chauhan's accomplishment is a reflection of his hard work, dedication, and unyielding passion for the field of veterinary science. This remarkable achievement not only underscores his academic brilliance but also cements his position as an inspiring figure in the veterinary community.

From an early age, Dr. Chauhan demonstrated an innate love for animals and a deep sense of responsibility toward their well-being. These qualities formed the foundation of his journey in veterinary science. His BVSc&AH (Honours) degree, completed with such distinction, is not just a testament to his intellectual capabilities but also to his tireless commitment to improving animal health and welfare. Pursuing a degree in Veterinary science is no small feat. It demands a unique combination of scientific expertise, practical skills, and emotional resilience. For Dr. Chauhan, the rigorous academic journey involved mastering a wide array of subjects, including animal Anatomy, Physiology, Pathology, Pharmacology, and Microbiology. Each subject posed its own set of challenges, yet he tackled them with unwavering determination and an insatiable thirst for knowledge. His exemplary performance, as evidenced by his impressive score of 80.3%, highlights his ability to excel even under pressure.

What sets Dr. Chauhan apart is not just his academic excellence but also his holistic approach to learning. Beyond textbooks and lectures, he actively sought opportunities to gain hands-on experience, volunteering at animal shelters and participating in community outreach programs. These experiences not only enriched his understanding of animal care but also shaped his perspective as a compassionate and empathetic veterinary professional.

Dr. Chauhan's achievement in earning his BVSc &AH (Honours) is particularly commendable in the context of the challenges he faced along the way. Balancing the demands of rigorous coursework with practical training is no easy task, yet he managed to excel in both areas. His ability to maintain such high standards of academic performance while also contributing to real-world veterinary practices speaks volumes about his time management skills, discipline, and work ethic.

One of the defining aspects of Dr. Chauhan's journey is his unwavering commitment to making a difference in the lives of animals and their owners. Throughout his academic career, he demonstrated a profound understanding of the critical role veterinarians play in society. From ensuring the health and welfare of pets to addressing public health concerns through the prevention of zoonotic diseases, Dr. Chauhan has always recognized the broader implications of his work. His academic success is thus not just a personal achievement but also a step forward in his mission to serve the community.

The journey to earning a BVSc &AH (Honours) is fraught with challenges, from long hours in laboratories to the pressure of exams and the constant need to stay updated with the latest advancements in veterinary science. Yet, Dr. Chauhan navigated these challenges with remarkable resilience and grace. His ability to consistently perform at such a high level is a testament to his mental fortitude and his passion for the field.

Dr. Chauhan's achievement is not just a reflection of his personal qualities but also a testament to the support and guidance he received from his mentors, peers, and family. Their encouragement played a crucial role in shaping his academic journey, and his success is a shared celebration of their collective efforts.

Looking ahead, Dr. Chauhan's BVSc &AH (Honours) degree was just the beginning of what promised to be a remarkable career in veterinary science. With his solid academic foundation, he was well-equipped to pursue advanced studies, contribute to cutting-edge research, or make a tangible difference through clinical practice. Whatever path he chose, one thing was certain: Dr. Chauhan continued to make significant contributions to the field of veterinary science, setting new benchmarks of excellence and inspiring others along the way.

As we celebrate this momentous occasion, it is important to recognize the broader significance of Dr. Chauhan's achievement. In a world where the bond between humans and animals is becoming increasingly important, professionals like Dr. Chauhan play a vital role in bridging the gap. Their work not only ensures the health and happiness of animals but also strengthens the human-animal connection, contributing to a more compassionate and harmonious society.

In conclusion, Dr. Chauhan's accomplishment in earning his BVSc &AH (Honours) with an outstanding 80.3% is a shining example of what can be achieved through hard work, perseverance, and a genuine passion for one's chosen field. His journey serves as an inspiration to aspiring veterinarians and a reminder of the transformative power of education. Dr. Chauhan, as you step into the next phase of your career, know that your achievements are a source of pride for all who know you. Your dedication, compassion, and brilliance are an inspiration to us all, and we look up to you in admiration on the remarkable impact you have undoubtedly made in the world of veterinary science.

Masters

Dr. Chauhan: A Journey of Passion and Excellence in Veterinary Pathology

Dr. Chauhan, a dedicated veterinarian and an exemplary student, recently achieved a milestone that underscores his unwavering passion for animal welfare and academic excellence. Completing his master's degree in veterinary science (Pathology) with an impressive 85.8%, Dr. Chauhan's journey is one of perseverance, hard work, and an unshakable commitment to making a difference in the lives of animals and their caretakers. Born and raised in a small town where agriculture and animal husbandry formed the backbone of the community, Ramswaroop's love for animals was evident from an early age. As a child, he would often accompany his father to the village veterinarian to treat the family's cattle. Observing the veterinarian's skill and compassion left a lasting impression on the young boy. "The way he calmed the animals and reassured my father inspired me. I knew I wanted to be like him," Ramswaroop recalls.

With limited resources but boundless determination, Dr Chauhan excelled in his studies, earning a spot at a reputed veterinary college for his undergraduate degree. It was during these formative years that he developed a deep understanding of Veterinary Pathology and minor as Microbiology. However, he was not content with just theoretical knowledge. He spent countless hours in the college's Veterinary hospital and Pathology laboratory, assisting seniors and learning the practical aspects of diagnosing and treating various ailments.

When the time came to pursue higher studies, Dr. Chauhan set his sights on a master's program that would allow him to specialize in Veterinary Pathology. "I wanted to do more than just treat animals; I wanted to give them a new lease on life," he explains. The journey to this dream, however, was far from easy. The admissions were tough, and the competition was fierce. Dr Chauhan's relentless efforts paid off, and he secured admission to one of the country's top Veterinary College at Pantnagar.

The master's program was both rigorous and rewarding. From advanced surgical techniques to the latest developments in animal healthcare, every aspect of the course deepened Dr Chauhan's knowledge and honed his skills. Yet, the challenges were immense. "Balancing academic demands with practical training was tough," he admits. "There were days when I would spend hours in the laboratory, followed by long nights of studying and doing practical."

Despite the demanding schedule, Dr. Chauhan made it a point to participate in community outreach programs organized by the institute. These programs involved providing free veterinary services in rural areas, where access to professional care was limited. "It was during these camps that I truly understood the plight of farmers and their animals," he says. "Their gratitude made every effort worthwhile."

His dedication to his studies and his hands-on approach did not go unnoticed. His professors often praised his analytical thinking and problem-solving abilities, particularly in complex cases. One such instance was when he successfully performed a life-saving diagnosis on a cow suffering from a twisted stomach. The case, which had baffled many, became a turning point in Dr. Chauhan's academic journey, earning him accolades from both his mentors and peers.

The master's program culminated in a comprehensive thesis, where Dr. Chauhan focused on developing diagnostic techniques for neonatal livestock. His research, which combined cutting-edge technology with practical applications, received widespread recognition. "I wanted my work to have a real-world impact," he says. "Seeing my research benefit animals and their owners was incredibly fulfilling." Scoring 85.8% in such a challenging program was no small feat, and he attributes his success to a combination of hard work, support from his family, and a genuine love for his field. "My parents were my biggest cheerleaders," he shares. "Their belief in me kept me going, even during the toughest times."

Now, as a qualified Veterinary Pathologist, Dr. Chauhan was eager to put his skills to use. His immediate plans included working in a state- of-the-art veterinary college, where he could continue to learn and grow. For aspiring veterinarians, Dr. Chauhan has a simple message: "Never lose sight of why you chose this profession. It's not just about treating animals; it's about understanding their needs and being their voice." He also emphasizes the importance of continuous learning. "Veterinary science is an ever- evolving field. Staying updated is crucial to providing the best care." Dr. Chauhan's story is a testament to the fact that passion, when combined with perseverance, can lead to extraordinary achievements. As he embarked on this new chapter of his life, there was little doubt that he would continue to excel, making a meaningful difference in the lives of countless animals and the people who depend on them.

In a world where the well-being of animals often takes a backseat, professionals like Dr. Chauhan serve as a beacon of hope. His journey is not just an inspiration for veterinary students but also a reminder of the profound impact that compassion and dedication can have in creating a better world for all living beings.

Dr. Chauhan's Groundbreaking Doctoral Research in Veterinary Science

Dr. Ramswaroop Chauhan, now celebrated as one of India's most accomplished veterinary scientists, began his journey toward excellence at G. B. Pant University of Agriculture & Technology, a premier institution known for producing some of the finest minds in agriculture and Veterinary sciences. His doctoral research at the university laid the foundation for a distinguished career in veterinary science, earning him national and international recognition. Enrolling in the doctoral program at G. B. Pant University in the late 1980s, Dr. Chauhan focused his research on a critical area of veterinary science: the epidemiology and control of bovine Rotaviral diarrhoea. This disease, a major challenge for dairy farmers globally, was responsible for significant economic losses in India as well. At a time when resources for advanced veterinary research were limited in India, Dr. Chauhan chose to tackle this complex issue with a combination of scientific rigor and innovative thinking.

The university, situated in Pantnagar, Uttarakhand, provided an ideal environment for Dr. Chauhan to delve into his research. The institution was well-regarded for its interdisciplinary approach, blending cutting- edge technology with practical agricultural applications. Dr. Chauhan was mentored by leading experts in veterinary microbiology, pathology, and epidemiology, who encouraged him to explore novel diagnostic techniques for bovine rotavirus infection. His doctoral research involved an intricate combination of fieldwork, laboratory experiments, and data analysis, making it one of the most comprehensive studies of its kind during that era.

Dr. Chauhan's work began with an extensive epidemiological survey across northern India. He visited many dairy farms to collect samples and assess the prevalence of calf diarrhoea rotavirus in different cattle breeds. His meticulous fieldwork not only provided valuable baseline data but also highlighted the socio-economic challenges faced by farmers in combating this disease. Dr. Chauhan's ability to connect with farmers and understand their practical difficulties added a unique dimension to his research, bridging the gap between scientific inquiry and real-world applications.

A significant aspect of Dr. Chauhan's doctoral thesis was the development of a rapid diagnostic test for early detection of rotavirus infection. During his research, he identified the limitations of traditional diagnostic methods, which were time-consuming, expensive, and often inaccessible to rural farmers. Drawing on his knowledge of microbiology and biochemistry, Dr. Chauhan developed an innovative field-based diagnostic kit that used enzyme-linked immunosorbent assay (ELISA) technology. This kit could detect early and subclinical infection—an asymptomatic form of the disease—at an early stage, enabling timely intervention.

Dr. Chauhan's diagnostic kit was a game- changer, receiving widespread acclaim from both his academic peers and veterinary practitioners. His method was not only cost- effective but also user-friendly, making it accessible to veterinarians and farmers alike. The results of this innovation were published in several prestigious international journals, such as Veterinary Record, Veterinary Microbiology, Journal of Comparative Pathology, bringing global attention to the work being conducted at G. B. Pant University.

Another groundbreaking aspect of Dr. Chauhan's doctoral research was his study on the role of cell mediated immunity in control of rotavirus infection in calves at local level. i.e. intestinal mucosal immunity; which paves the way for development of effective oral vaccine for the control of rotavirus infection in animals and in human beings. His recommendations formed the basis for several policy discussions on antimicrobial stewardship in livestock management.

In addition to his diagnostic and antimicrobial studies, Dr. Chauhan explored the immunological aspects of bovine neonatal diarrhoea. His experiments demonstrated how certain host factors, such as genetic predisposition and immune responses, influenced a cow's susceptibility to infection. These insights opened up new avenues for selective breeding programs aimed at developing disease- resistant cattle breeds.

Dr. Chauhan's commitment to excellence was evident in the numerous accolades he received during his doctoral program. His thesis, titled Epidemiological and Immunological Studies on neonatal bovine diarrhoea with Special Reference to Diagnostic and Therapeutic Interventions, was awarded the university's Best Research Thesis Award, a recognition given to only the most outstanding scholarly contributions. Additionally, he presented his findings at various national and international conferences, earning accolades for his clarity of thought and innovative approaches.

The impact of Dr. Chauhan's doctoral research extended far beyond the academic sphere. His diagnostic kit was later commercialized and adopted by veterinary practitioners across India, significantly improving the management of rotavirus infection. Dr. Chauhan often credits the supportive environment at G. B. Pant University for his success. The institution's emphasis on practical, solution-oriented research aligned perfectly with his vision of veterinary science as a tool for improving the livelihoods of farmers. He frequently acknowledges the guidance of his mentors, whose expertise and encouragement played a crucial role in shaping his academic journey.

Reflecting on his doctoral days, Dr. Chauhan describes them as the most intellectually stimulating period of his life. He recalls the long hours spent in the laboratory, the countless farm visits, and the lively discussions with his peers and professors. These experiences not only enriched his understanding of veterinary science but also instilled in him a lifelong commitment to research and innovation.

Today, Dr. Chauhan's doctoral work at G. B. Pant University is regarded as a milestone in the field of veterinary science. His ability to integrate advanced research with practical applications has inspired a generation of veterinarians and scientists. As he often says, "The purpose of veterinary science is not just to cure animals but to create a sustainable ecosystem where both animals and humans can thrive. My journey began at Pantnagar, and it is a legacy I hope to pass on to future scholars."

7

A Trailblazer in Veterinary Science Celebrated Journey of Excellence

Dr. Chauhan, a renowned figure in the field of veterinary science, has left an indelible mark on the discipline through his groundbreaking research, innovative teaching, and unwavering dedication to the advancement of animal health and welfare. Over the years, his exceptional contributions have earned him a multitude of prestigious awards, medals, and honors, making him a luminary in his field.

His illustrious career began to gain recognition with the Best Young Scientist Award in 1992by Indian Association of Veterinary Pathologists, an acknowledgment of his potential and early contributions to veterinary science in general and Veterinary Pathology in particular. This award marked the start of a journey filled with accolades that would underscore his relentless pursuit of excellence. Four years later, in 1996, he was honored with the IAAVR Award, further solidifying his position as a promising researcher and academic.

By 1999, Dr. Chauhan's expertise and dedication had earned him the National Fellow Professor chair of ICAR, a prestigious recognition that highlighted his growing influence in veterinary research. Around the same time, he was conferred the title of Fellow NAVS (2000) and Fellow SIIP (2001), reflecting his significant contributions to national and international veterinary organizations. These fellowships were a testament to his role in advancing scientific understanding and his collaborative efforts within the veterinary community.

Dr. Chauhan's profound impact on the field was also recognized with several specialized awards, including the K.S. Nair Memorial Award in 1999, the Vigyan Bharti Award in 2000, and the Dr. C.M. Singh Trust Award in 2002. These honors celebrated his ability to merge academic rigor with practical applications, benefiting both the academic community and the agricultural sector. Similarly, the Dr. Rajendra Prasad Award (2002) and the Shri Ramlal Agrawal National Award (2000) underscored his contributions to veterinary science in India.

One of Dr. Chauhan's hallmark achievements was his consistent excellence in publishing high-impact research papers. His groundbreaking work earned him the Best Paper Award SIIP (2003) and a streak of victories in the Best Paper Award JIIP (2001, 2002, 2003) competitions. These accolades highlighted not only the innovative nature of his research but also its relevance and utility to the veterinary profession.

As an educator, Dr. Chauhan's passion for teaching and mentoring was recognized with the Best Teacher Award (2004) by GBPUAT. This honor reflected his ability to inspire and shape the next generation of veterinary scientists, fostering a culture of curiosity and academic excellence. His contributions to veterinary pathology also earned him the title of Fellow IAVP (2006), further establishing his authority in the field. Dr. Chauhan's influence extended beyond academia, as evidenced by awards such as Gopal Gaurav (2007) and the Bharat Excellence Award (2007). These recognitions celebrated his role in promoting animal health and welfare at the national level, showcasing his ability to blend scientific innovation with societal impact.

In 2008, Dr. Chauhan reached another milestone in his career by receiving the Diplomat, ICVP, and the Intas-ISVE Best Veterinary Scientist Award. These accolades not only recognized his technical expertise but also his leadership in shaping the future of veterinary pathology. His recognition as the Best Academician Award (2012) highlighted his ability to balance research, teaching, and administrative responsibilities with remarkable efficiency.

Dr. Chauhan's contributions to traditional and alternative veterinary medicine were also recognized with the Moropant Pingle Go Sewa National Award (2015). This award acknowledged his dedication to promoting indigenous practices and integrating them with modern veterinary science. His groundbreaking achievements were further celebrated with the Outstanding Scientist Award (2019) and the Research Excellence Award (2020), both of which underscored his enduring commitment to advancing the frontiers of knowledge in veterinary science.

In 2020, Dr. Chauhan was honored with the Indo Asian-Claude Bourgelat Distinguished Innovative Scientist Award in Animal Immunopathology, a testament to his pioneering research in this specialized field. This award symbolized international recognition of his innovative approaches and contributions to understanding animal immune systems. In the same year, he earned the title of MAN OF COWPATHY at GADVASU Ludhiana, a unique honor that celebrated his work in promoting the health and productivity of cattle through innovative veterinary practices. Throughout his career, Dr. Chauhan's accolades have not only celebrated his individual achievements but have also highlighted the broader impact of his work on society. His ability to bridge the gap between scientific research and practical applications has made him a true pioneer in the field. Each award and honor serves as a milestone in his journey, reflecting his dedication to excellence and his vision for a healthier and more sustainable future for animals and humans alike.

Dr. Chauhan's decorated career is a testament to his tireless efforts and unwavering passion for veterinary science. His numerous awards and honors are not just a recognition of his individual accomplishments but also an inspiration to countless others in the field. By setting new benchmarks of excellence, Dr. Chauhan has established himself as a beacon of hope and progress in the world of veterinary medicine.

Recognition

Dr. Chauhan is a distinguished scientist and academician who has made significant contributions to the fields of veterinary pathology, immunology, and environmental health. Over the years, he has held numerous prestigious positions in national and international committees, advisory boards, and research organizations. His expertise and dedication have earned him recognition across a wide spectrum of scientific domains.

Key Appointments and Roles

Dr. Chauhan has been actively involved in shaping policies, advancing research, and mentoring young scientists. His national-level contributions include being a member of the Hindi Advisory Committee of the Government of India, a role where he promoted the use of Hindi in scientific communications. Internationally, he has served as a member of the WHO/IPCS Committee on Environmental Health Criteria, where he contributed to global discussions on environmental and public health.

In India, Dr. Chauhan has chaired several influential committees. He led the Indian Association of Veterinary Pathologists (IAVP) and the Veterinary Council of India (VCI) in revising and modernizing veterinary pathology courses, ensuring they align with global standards. He also chaired the Cow Therapy Research Group, reflecting his deep interest in traditional and innovative veterinary practices.

He was part of the Central Management Committee of AIIMS (All India Institute of Medical Sciences), where his inputs shaped healthcare policies. Additionally, he contributed to the Board of Management at the Indian Veterinary Research Institute (IVRI) and served on the Executive Council of the National Research Institute (NRI). His leadership extended to various technical selection committees, where he served as chairman or member, including the Agricultural Scientists Recruitment Board (ASRB), Central Agricultural Universities (CAUs), and State Agricultural Universities (SAUs).

Dr. Chauhan is also a recognized technical expert and reviewer for CAB International, a leading intergovernmental organization in agricultural and environmental sciences. Furthermore, he has provided peer reviews for several renowned national and international journals, demonstrating his profound impact on the academic community.

Research Contributions and Scientific Breakthroughs

Dr. Chauhan has spearheaded numerous groundbreaking research projects, many of which have transformed the field of veterinary science. He added to the understanding of critical diseases such as enterotoxaemia, pyometra, and ETEC infection in camels. His work on isolating the camelpox virus was a first in India and contributed significantly to disease management in camel populations. One of his notable achievements includes the development of the Dot Immunobinding Assay (DIA), a rapid diagnostic test that revolutionized disease detection in India. He also elucidated the role of cell-mediated immunity in combating rotavirus infections in calves, a pioneering study that deepened the understanding of immunological responses in young livestock.

Dr. Chauhan has made significant strides in the study of immunopathology caused by environmental and chemical agents. His research on the toxic effects of pesticides, heavy metals, mycotoxins, and nanoparticles has provided insights into their impact on animal and human health. He also developed a novel method using MTT dye to detect cell- mediated immunity (CMI) responses, a technique now widely adopted in immunological studies.

His innovative research on Panchgavya, a traditional Indian concoction derived from cow products, led to its scientific validation. Dr. Chauhan coined the term "Cowpathy", which represents the therapeutic applications of Panchgavya. This work bridges the gap between traditional knowledge and modern science, offering new avenues for alternative medicine.

International Recognition and Collaborations

Dr. Chauhan's expertise has earned him global recognition. He was invited as a visiting professor at the University of Wageningen in the Netherlands, where he collaborated with leading researchers on veterinary and environmental health topics. He also served as an advisor to the World Health Organization (WHO), contributing to global health initiatives. In addition to his academic roles, Dr. Chauhan has held editorial responsibilities. He served as the Managing Editor of the Journal of Immunology and Immunopathology and the Editor-in-Chief of the International Journal of Cow Science. These roles allowed him to shape the dissemination of scientific knowledge and foster collaboration among researchers worldwide.

Conferences, Workshops, and Training Programs

Dr. Chauhan has a strong commitment to knowledge sharing and capacity building. He has organized several high-profile conferences and training programs to address pressing scientific challenges. Notable among these are:

The II SIP Convention on Immunomodulation in Health and Disease, which brought together experts to discuss the role of the immune system in disease prevention and treatment. A brainstorming session on tuberculosis (TB), Johne's disease (JD), and sexually transmitted diseases (STDs) in bovines, aimed at improving livestock health and productivity. The International Association of Veterinary Pathologists (IAVP) Conference at IVRI and the 5th Society for Immunology and Immunopathology (SIP) Conference.

Dr. Chauhan has also organized several e- conferences, including the International Conference on Immunology in the 21st Century, which focused on advancing one- health approaches—a concept that links human, animal, and environmental health. To mentor the next generation of scientists, Dr. Chauhan has conducted over 40 short-term training programs from 2009 to 2021. These programs covered diverse topics such as immunopathology (1998), Dot Immunobinding Assay (2001), Cell Stress and Apoptosis (2001), Molecular Diagnostic Techniques (2004), and the Art of Disease Investigation (2006-2008). His training courses have equipped young scientists with cutting-edge skills and techniques, enhancing their research capabilities.

Research Funding and Projects

Dr. Chauhan has been the principal investigator (PI) of 25 research projects, with funding worth millions of rupees. These projects were supported by prominent organizations such as the Indian Council of Agricultural Research (ICAR), the Department of Science and Technology (DST), the UP Council for Agricultural Research (UPCAR), and the Department of Animal Husbandry and Dairying (DADF). His collaborations with industry partners have further strengthened the application of research findings in real-world scenarios.

Publications and Academic Legacy

Dr. Chauhan has authored numerous research papers, book chapters, and reviews in reputed national and international journals. His publications have been widely cited, reflecting the impact of his work on the global scientific community. Through his writings, he has addressed critical issues in veterinary science, immunology, and environmental health, providing valuable insights and solutions. His academic legacy extends beyond his research contributions.

As a mentor, he has guided countless students and researchers, many of whom have gone on to become leaders in their respective fields. His emphasis on scientific rigor, innovation, and ethical practices has inspired a new generation of scientists.

Conclusion

Dr. Chauhan's illustrious career is a testament to his unwavering dedication to advancing science and improving public health. His contributions span a wide range of disciplines, from veterinary pathology and immunology to environmental health and traditional medicine. Through his research, leadership, and mentorship, he has left an indelible mark on the scientific community, both in India and abroad. With a career characterized by innovation, collaboration, and service, Dr. Chauhan continues to be a source of inspiration for aspiring scientists. His work not only addresses current challenges but also lays the foundation for future advancements in health, agriculture, and environmental sustainability.

8

The Legacy of Dr. Chauhan: A Journey of Triumph and Compassion

Dr. Chauhan's illustrious career as a Veterinary Pathologist is marked by resilience, unwavering compassion, and a commitment to animal welfare. Despite facing numerous challenges, his journey serves as an inspiration to many, not only for his professional achievements but also for his ability to turn obstacles into stepping stones for greater success. Over decades of practice, Dr. Chauhan became a symbol of hope for animals and their caregivers, a testament to what dedication and empathy can accomplish.

Breakthroughs and Milestones

Dr. Chauhan's career is filled with milestones that showcase his determination to push the boundaries of veterinary medicine. Early in his career, he made headlines by successfully treating a prized racehorse suffering from a rare neurological disorder. The case, which had been deemed incurable by others, required not only medical expertise but also the courage to think outside the box. Dr. Chauhan used a combination of experimental techniques and traditional therapies to restore the horse to full health. This triumph not only saved the horse's career but also cemented Dr. Chauhan's reputation as an innovative problem-solver.

Another significant breakthrough came when Dr. Chauhan devised a novel treatment protocol for canine distemper, a disease that ravages stray dog populations. By combining immunotherapy and supportive care, he achieved remarkable survival rates, helping communities across the region manage outbreaks more effectively. One of his most celebrated cases involved Tara, a critically injured elephant rescued from an illegal logging operation. Tara had sustained multiple fractures and severe infections, and her chances of survival were slim. Dr. Chauhan spearheaded an intensive six-month treatment regimen, which included groundbreaking orthopedic surgery and customized physiotherapy. Tara's recovery became a symbol of hope for wildlife conservationists and sparked a nationwide discussion about the ethics of using animals in hazardous labor.

Dr. Chauhan's ability to combine advanced medical techniques with empathy and innovation earned him widespread recognition. Over time, his clinic transformed into a state- of-the-art facility, drawing clients from across the country and even neighboring regions.

Advocacy and Education

Dr. Chauhan's contributions extended far beyond his clinic walls. He was a tireless advocate for animal welfare, using his platform to influence policies and educate communities about the humane treatment of animals. His advocacy efforts came to the forefront during a rabies outbreak in a neighboring district. When misinformation led to the mass culling of stray dogs, Dr. Chauhan intervened, proposing a comprehensive vaccination and sterilization program instead. His approach not only controlled the outbreak but also demonstrated that humane solutions are both effective and sustainable. Recognizing the power of education, Dr. Chauhan launched workshops and seminars for aspiring veterinarians. His teaching extended to rural areas, where he trained local communities in basic animal care and first aid. This grassroots initiative reduced preventable animal deaths and improved public perception of animal welfare. A major milestone in his educational efforts was the establishment of a scholarship fund for underprivileged students pursuing veterinary studies. By addressing financial barriers, Dr. Chauhan ensured that talented individuals could enter the field, irrespective of their socio- economic background.

Contributing to Wildlife Conservation

Dr. Chauhan's passion for wildlife conservation was another defining aspect of his career. He collaborated with national parks and sanctuaries to rehabilitate injured animals and ensure their safe return to the wild. His work often involved species on the brink of extinction, where every life saved made a significant impact. A particularly memorable case was that of a Himalayan snow leopard suffering from a severe infection. The snow leopard, a key predator in the region's ecosystem, had been injured in a poacher's trap. Dr. Chauhan's meticulous treatment plan, which included advanced surgical techniques and infection management, saved the animal's life. The snow leopard's recovery was hailed as a victory for conservationists, drawing attention to the urgent need to combat poaching.

Dr. Chauhan also played a pivotal role in efforts to save the Indian pangolin, a species threatened by illegal wildlife trade. He worked alongside international researchers to develop rehabilitation protocols, ensuring that rescued pangolins could adapt to the wild after their release. His dedication to wildlife often required him to work under challenging conditions, from remote jungle camps to high-altitude terrains. These efforts not only saved countless animal lives but also contributed valuable research to the global conservation community.

Innovations in Veterinary Medicine

Innovation was a hallmark of Dr. Chauhan's practice. He introduced techniques that revolutionized veterinary medicine in his region, such as minimally invasive surgeries and the use of advanced imaging technologies for diagnostics. His research into herbal remedies for chronic conditions in livestock proved to be a game-changer for rural farmers, who struggled with the high costs of conventional treatments.

Dr. Chauhan's clinic became a hub for experimentation and learning, attracting veterinarians from across the world to collaborate on complex cases. One such collaboration led to the development of a groundbreaking treatment for a viral disease affecting big cats in captivity. This research not only saved the lives of numerous tigers and lions but also highlighted the importance of cross-border knowledge-sharing in veterinary science.

Global Recognition

Dr. Chauhan's achievements earned him invitations to international veterinary conferences, where he shared his experiences and innovations. His work with endangered species and his efforts to combat zoonotic diseases (Diseases transmissible from animals to man) gained particular attention. He was recognized with multiple awards, including a prestigious international honor for his lifetime contributions to veterinary medicine and wildlife conservation.

Despite his global recognition, Dr. Chauhan remained grounded, often emphasizing the importance of local action. He believed that meaningful change began at the community level, where simple acts of compassion could create ripple effects.

A Career Defined by Compassion

Compassion was the cornerstone of Dr. Chauhan's career. He believed that every animal, regardless of size or species, deserved care and dignity. This philosophy was evident in his hands-on approach, whether it was spending sleepless nights monitoring critical patients or traveling long distances to attend to an injured animal in a remote area. One of the most heartwarming moments in his career was reuniting a severely injured pet dog with its owner. The dog had gone missing for months and was brought to Dr. Chauhan's clinic in critical condition. After weeks of intensive care, the dog made a full recovery, and the emotional reunion between pet and owner became a testament to the transformative power of veterinary medicine.

Dr. Chauhan also focused on the mental health challenges faced by veterinarians. Aware of the emotional toll the profession could take, he founded support groups and helplines for his peers. This initiative created a sense of community among veterinarians, fostering resilience and reducing burnout in the field.

Legacy and Lessons

Dr. Chauhan's legacy is one of perseverance, empathy, and unwavering dedication to the well-being of animals. His journey from a small-town practitioner to a globally respected veterinarian serves as an inspiration to professionals across disciplines. As he reflects on his career, Dr. Chauhan takes pride not only in his accomplishments but also in the lives he has touched. His story is a powerful reminder that challenges, no matter how daunting, can be overcome with determination and a clear sense of purpose.

Dr. Chauhan's words continue to resonate: "Every life matters, and every effort to save a life, no matter how small, is worth it. Compassion is the greatest medicine we have, and it is limitless."

9

Personal Philosophy

Dr. Chauhan's philosophy as a Veterinary Pathologist reflects a profound understanding of the interconnectedness of life and an unwavering commitment to the well-being of animals. His approach is rooted in compassion, scientific rigor, and a holistic appreciation of the delicate balance that sustains ecosystems. Over the years, he has become a beacon of hope for gaushalas and cow enthusiasts, demonstrating that veterinary medicine is not just a profession but a calling that demands empathy, dedication, and an unyielding respect for life.

At the core of Dr. Chauhan's philosophy is the belief that every living being, regardless of its size or species, deserves dignity and care. To him, animals are not mere creatures but sentient beings with emotions, instincts, and intrinsic value. This perspective shapes his practice, as he treats each animal with the same care and attention he would give to a human patient. Whether it's a stray dog in need of emergency care, a farm animal requiring preventive medicine, or an endangered species in a conservation program, Dr. Chauhan approaches his work with equal seriousness and respect.

Dr. Chauhan emphasizes the importance of understanding animals within the context of their natural behaviors and environments. He often says, "To heal an animal, you must first listen to what it cannot say." This insight has driven him to study not only veterinary medicine but also animal behavior, ecology, and the complex relationships between humans and animals. He believes that understanding these dynamics is essential for providing comprehensive care. For example, he advocates for preventive care practices that align with an animal's natural rhythms, such as vaccination schedules that account for seasonal changes and diets that mimic an animal's natural food sources.

A key element of Dr. Chauhan's philosophy is education. He is a staunch advocate for spreading awareness about animal welfare and responsible pet ownership. He believes that much of the suffering animals endure stems from a lack of knowledge or misguided practices. Through workshops, community outreach programs, and collaborations with Animal Welfare Board of India, he has worked tirelessly to instill a sense of responsibility in people, particularly young minds. Dr. Chauhan often tells his students, "The health of animals is intertwined with the health of our planet and ourselves. When we care for them, we are caring for our future."

Another cornerstone of Dr. Chauhan's approach is his commitment to innovation and lifelong learning. He firmly believes that the field of veterinary medicine is ever-evolving, and staying abreast of new research and technologies is not optional but a moral obligation. He has invested considerable time in studying advanced diagnostic techniques, minimally invasive surgical methods, and alternative therapies like cowpathy and herbal medicine. His integrative approach combines the best of traditional veterinary science with emerging practices, ensuring that his patients receive the most effective and humane care possible. Dr. Chauhan also places great importance on the mental and emotional well-being of animals. He recognizes that stress and anxiety can exacerbate physical ailments, and he strives to create a calming environment in his clinic. From using pheromone diffusers to employing gentle handling techniques, he ensures that his patients feel safe and cared for. His emphasis on emotional well-being extends to the animals' owners as well. Dr. Chauhan believes that empathy for the pet owner is as crucial as empathy for the animal. He often takes the time to explain medical conditions and treatment options in simple terms, ensuring that families feel informed and supported during challenging times.

One of the most remarkable aspects of Dr. Chauhan's philosophy is his dedication to wildlife conservation. He views his work as a small but significant contribution to the preservation of biodiversity. Over the years, he has participated in several rescue and rehabilitation projects, often working in challenging conditions to save injured or orphaned wildlife. His efforts have not only saved countless lives but have also inspired others to take up the cause of conservation. Dr. Chauhan believes that humanity's survival is intrinsically linked to the survival of other species, and he is determined to protect that delicate balance.

Dr. Chauhan's philosophy also extends to the ethics of veterinary practice. He is deeply committed to the principle of "do no harm," even in difficult circumstances. He has often been faced with tough decisions, such as whether to euthanize a suffering animal or attempt risky treatments. In these moments, he relies on a combination of scientific evidence, ethical reasoning, and empathy to guide his choices. His unwavering commitment to ethical practice has earned him the respect of his peers and the trust of his clients.

In essence, Dr. Chauhan's philosophy is a testament to the power of compassion, knowledge, and responsibility. He believes that every action, no matter how small, can make a difference in the life of an animal and, by extension, the world. His work is not just about healing bodies but also about fostering a deeper connection between humans and animals, one that is rooted in mutual respect and care. Through his philosophy and practice, Dr. Chauhan has shown that being a veterinarian is not just a job; it is a way of life, a commitment to making the world a better place for all living beings.

The Personal Touch: Dr. Chauhan's Approach to cow Care

Dr. RS Chauhan's name was synonymous with trust in the world of cow care. Known for his calm demeanor and meticulous attention to detail, he brought not only skill but also a deeply personal approach to every case he handled. For Dr. Chauhan, veterinary medicine wasn't just about treating an ailment—it was about forging connections, understanding unspoken needs, and creating a space where both animals and their humans felt safe and valued.

Listening Without Words

One of the most striking aspects of Dr. Chauhan's approach was his ability to "listen" to animals. He believed that every creature, no matter how small or large, communicated its needs in its own way. Whether it was the subtle change in a dog's posture, the restless shifting of a cow, or the distant gaze of a bird, Dr. Chauhan paid attention to these cues. "Animals can't tell you what hurts, but they always show you," he would say.

He would often spend extra time observing his patients before conducting physical examinations. This habit of careful observation allowed him to pick up on symptoms that others might miss. For example, he once correctly identified an uncommon parasitic infection in a cow simply by noticing its unusual breathing pattern, something even the owner hadn't detected.

Holistic Care and Emotional Sensitivity

Dr. Chauhan's approach went beyond physical health; he always considered the emotional well-being of the animals. He believed stress and anxiety could worsen physical ailments, so he took care to create a calm and reassuring environment in the gaushalas. He spoke in soft tones, maintained gentle eye contact, and often kept a hand on the animal to convey reassurance. Dr. Chauhan knew that these small gestures made a world of difference. "An animal that trusts you is an animal that heals faster," he would say.

Personalized Care Plans

Unlike many veterinarians who relied on a one- size-fits-all approach, Dr. Chauhan believed in tailoring his treatments to the unique needs of each patient. He took into account not only the animal's medical history but also its lifestyle, diet, and even its personality. "Health isn't just about medicine—it's about balance," he explained. Dr. Chauhan also extended this personalized approach to educating animal owners. Instead of overwhelming them with technical jargon, he broke down complex conditions into simple, relatable terms. He encouraged owners to ask questions, even if they seemed trivial, and took the time to ensure they felt confident in managing their pet's care at home.

Building Trust with Owners

Dr. Chauhan understood that behind every animal was an owner who was often anxious, overwhelmed, or even grieving. He made it a point to address their concerns with empathy and patience. Whether it was a worried farmer bringing in a sick calf or a first-time pet owner with a kitten, Dr. Chauhan treated every case with the same level of care and respect; he knows how to put my mind at ease." This reputation for building trust wasn't just a byproduct of his skills; it was a deliberate effort on his part to ensure that every client felt supported and understood.

Proactive Over Reactive

Dr. Chauhan's approach was rooted in prevention. He believed that the best way to ensure an animal's well-being was to prevent illness before it occurred. To this end, he ran regular vaccination drives, held free workshops on proper animal care, and even created detailed seasonal care guides for farmers. "An ounce of prevention is worth a pound of cure—for both the animal and the owner," he often quipped. He also encouraged routine check-ups, even for seemingly healthy animals. His proactive methods earned him the trust of countless families, farmers, and even wildlife conservationists who sought his guidance.

The Gentle Goodbye

One of the most difficult aspects of Dr. Chauhan's practice was handling end-of-life care. When an animal was beyond saving, he approached the situation with unmatched grace and sensitivity. He ensured that the process was as painless and peaceful as possible, both for the animal and its owner. He often sat with grieving families, offering them quiet words of comfort and reminding them of the joy their pet had brought into their lives. "I don't just help animals live well," he once said. "I also help them leave this world with dignity."

Legacy of Care

Dr. Chauhan's personal approach wasn't just about technical expertise—it was about the relationships he built along the way. He saw his role as a bridge between humans and animals, fostering a deeper understanding of the bond they shared. His unwavering commitment to compassion, respect, and individualized care left an indelible mark on everyone who walked through his gauma farm doors. In the words of one of his long-time clients, “Dr. Chauhan doesn’t just treat animals. He heals hearts, restores hope, and reminds us all of the beauty in caring for those who cannot speak.

10

Future Goals of Dr. Chauhan in Expanding the Scope of Veterinary Science

As he looks to the future, Dr. Chauhan envisions several key goals to further the field of veterinary medicine: Expanding the scope of Veterinary Science :

Future goals of Dr. Chauhan

1. **Advancing Research in Veterinary Pathology:** Building upon his extensive research background, which includes authoring 120 books, contributing 101 chapters, and publishing 239 research papers, Dr. Chauhan aims to delve deeper into emerging diseases affecting livestock and companion animals. He plans to focus on molecular pathology to better understand disease mechanisms at the genetic level, facilitating the development of targeted treatments and preventive measures.
2. **Promoting 'Cowpathy' and Indigenous Practices:** Known as the "Cowpathy Man," Dr. Chauhan has been a proponent of integrating traditional knowledge with modern veterinary practices. He intends to conduct rigorous scientific studies to validate and promote the use of indigenous products in animal healthcare, aiming to offer sustainable and eco-friendly alternatives to conventional treatments.
3. **Enhancing Veterinary Education:** With a passion for veterinary education, Dr. Chauhan plans to develop comprehensive curricula that incorporate the latest advancements in veterinary science. He seeks to mentor the next generation of veterinarians, emphasizing the importance of ethical practices, continuous learning, and adaptability in the face of evolving challenges in animal health.
4. **Strengthening One Health Initiatives:** Recognizing the inter-connectedness of human, animal, and environmental health, Dr. Chauhan aims to collaborate with interdisciplinary teams to address zoonotic diseases. By contributing to One Health initiatives, he hopes to develop

strategies that mitigate the risks of disease transmission between animals and humans, ensuring a holistic approach to global health.

5. **Advocating for Animal Welfare:** As a member of the Animal Welfare Board of India and the Committee for the Purpose of Control and Supervision of Experiments on Animals (CPCSEA), Dr. Chauhan is committed to promoting humane treatment of animals. He plans to work towards strengthening animal welfare regulations, increasing awareness about ethical treatment, and ensuring the implementation of best practices in animal care.
6. **Leveraging Technology in Veterinary Practice**: Embracing the digital age and big data, Dr. Chauhan envisions the integration of technology in veterinary diagnostics and treatment. He aims to explore the use of telepathology and telemedicine to reach underserved areas, develop digital tools for disease surveillance, and utilize data analytics to inform decision-making in animal health management.
7. **Global Collaboration and Knowledge Exchange**: Understanding the value of global perspectives, Dr. Chauhan plans to foster international collaborations to share knowledge, research findings, and best practices. By participating in global conferences and contributing to international journals, he seeks to position Indian veterinary research on the world stage, facilitating mutual growth and learning.
8. **Environmental Conservation & Sustainable Practices:** Acknowledging the impact of environmental factors on animal health, Dr. Chauhan aims to advocate for sustainable agricultural and veterinary practices. He plans to research the effects of climate change on livestock diseases and develop strategies to mitigate these impacts, ensuring the resilience of animal agriculture in changing environmental conditions.
9. **Public Health and Food Safety**: With a focus on public health, Dr. Chauhan intends to work on improving food safety standards related to animal products. He aims to develop better diagnostic tools for detecting pathogens in food, establish guidelines for antibiotic use in livestock to combat resistance, and educate farmers and the public on best practices to ensure the safety of the food supply.
10. **Mentorship and Capacity Building**: Committed to building capacity in the veterinary field, Dr. Chauhan plans to establish mentorship programs for young veterinarians and researchers. He aims to provide guidance, resources, and opportunities for professional development, ensuring a robust pipeline of skilled professionals dedicated to advancing animal health and welfare.

Through these comprehensive goals, Dr. R.S. Chauhan seeks to continue his lifelong dedication to veterinary science, integrating traditional wisdom with modern innovations to address the evolving challenges in animal health.

11. **Developing Advanced Diagnostic Technique**s: One of Dr. Chauhan's key objectives is to enhance disease diagnostics in veterinary medicine. With his expertise in pathology and molecular biology, he envisions developing rapid, cost-effective diagnostic tools that can detect emerging animal diseases early. He aims to collaborate with biotech companies and research institutions to create portable diagnostic kits that can be used in remote locations, improving disease surveillance and outbreak management.
12. **Encouraging Entrepreneurship in Veterinary Science**: Dr. Chauhan believes that veterinarians can play a crucial role in the economy by setting up animal healthcare startups, pet care clinics, and pharmaceutical research firms. He plans to support budding veterinary entrepreneurs by organizing training programs, workshops, and providing mentorship in business development. His goal is to bridge the gap between research and commercialization, ensuring that innovative veterinary solutions reach the market effectively.
13. **Strengthening Veterinary Policies and Legal Frameworks**: As an advocate for policy reforms, Dr. Chauhan is committed to influencing government policies on livestock health, food safety, and veterinary education. He aims to work closely with policymakers to ensures stricter implementation of animal welfare laws, improved veterinary healthcare infrastructure, and better funding for veterinary research. His vision includes lobbying for increased financial support for rural veterinary services to make quality healthcare accessible to livestock farmers.
14. **Popularizing Veterinary Science Among Youth**: Dr. Chauhan understands that the future of veterinary medicine lies in inspiring young minds. He plans to introduce awareness programs in schools and colleges to educate students about the importance of veterinary science, encouraging them to consider it as a viable career option. Through public lectures, social media outreach, and interactive learning modules, he hopes to attract more talented individuals to the field.
15. **Writing and Knowledge Dissemination**: Having authored over 100 books and contributed extensively to academic literature, Dr. Chauhan intends to continue his legacy of knowledge-sharing. He plans to write more books focusing on modern veterinary challenges, disease

management, and indigenous animal healthcare solutions. His goal is to create comprehensive, easily accessible learning materials for veterinary students, practitioners, and animal health enthusiasts.

16. **Animal-Assisted Therapy and Human- Animal Bond Research**: Dr. Chauhan also envisions promoting the benefits of animal-assisted therapy for mental health and rehabilitation. By conducting research on the positive impacts of animal interaction on human well-being, he aims to advocate for the inclusion of therapy animals in hospitals, rehabilitation centers, and special- needs schools. His future research could pave the way for integrating animal therapy into mainstream medical treatment.

17. **Global Impact and Recognition:** With India becoming a key player in global veterinary science, Dr. Chauhan seeks to strengthen India's position in international veterinary research. He aims to establish research collaborations with veterinary institutions worldwide, fostering an exchange of knowledge and technology. By contributing to global veterinary forums, he hopes to represent Indian veterinary advancements on a broader platform.

Through these expanded goals, Dr. Chauhan continues to shape the future of veterinary medicine, ensuring that his work benefits not only animals but also humanity and the environment. His relentless dedication to research, education, and animal welfare cements his place as a visionary in the field.

11

The Community Engagements of Veterinary Scientist Prof Chauhan in Animal Welfare Activities

Dr. Chauhan, a dedicated veterinary scientist, has built a remarkable reputation for his tireless efforts in animal welfare. His journey into veterinary medicine was not just a career choice but a calling, deeply rooted in his compassion for animals and a firm belief in their right to a healthy, dignified life. Over the years, he has not only saved countless animals but also played a crucial role in shaping community attitudes towards animal welfare. His engagements span various domains, including free medical camps, rescue missions, awareness campaigns, collaboration with NGOs, and advocacy for animal rights.

Free Veterinary Camps in Rural and Urban Areas

One of Dr. Chauhan's most impactful contributions has been his organization of free veterinary camps. These camps, held in both rural and urban areas, provide essential medical aid to animals that might otherwise go untreated. Many pet owners and farmers struggle with the financial burden of veterinary care, and Dr. Chauhan's initiative ensures that their animals receive timely vaccinations, deworming, and treatment for common illnesses.

In rural areas, where livestock plays a critical role in the livelihoods of farmers, these camps have been particularly beneficial. By treating cattle, goats, and poultry, Dr. Chauhan has helped improve the overall health of farm animals, leading to better productivity and economic stability for local communities. In urban areas, his focus extends to street animals, offering free treatments for injured or ailing stray dogs and cats.

Street Animal Rescue and Rehabilitation

Dr. Chauhan has been instrumental in rescuing and rehabilitating street animals, particularly those that have suffered due to accidents, malnutrition, or human cruelty. Working closely with animal rescue organizations, he ensures that injured animals receive prompt medical attention. His clinic has often doubled as a temporary shelter for rescued animals, where they are nursed back to health before being placed in permanent homes.

One of his notable rescue missions involved a stray dog that had been severely injured in a road accident. Bystanders had ignored the suffering animal until Dr. Chauhan stepped in, performing emergency surgery and providing post-operative care. After months of rehabilitation, the dog was adopted by a loving family. Stories like these have not only saved lives but have also inspired others to be more empathetic toward animals in distress.

Public Awareness and Educational Initiatives

Understanding that long-term change requires educating the public, Dr. Chauhan actively conducts awareness campaigns about responsible pet ownership, the importance of vaccinations, and ethical treatment of animals. He visits schools, colleges, and community centers, engaging with students and residents to instill a sense of responsibility toward animals. One of his key messages revolves around the importance of sterilization to control the stray population humanely. He explains how unchecked breeding leads to suffering, overcrowding in shelters, and increased cases of animal abandonment. Through his workshops, he has encouraged many communities to **participate in sterilization programs and vaccination drives.**

Additionally, Dr. Chauhan uses social media as a powerful tool to spread awareness. Through informative posts, live Q&A sessions, and success stories of rescued animals, he reaches a broad audience, encouraging more people to contribute to animal welfare.

Collaboration with NGOs and Government Bodies

Dr. Chauhan has formed strong partnerships with various NGOs and government bodies to amplify his impact. He collaborates with organizations that focus on animal rights, ensuring that laws protecting animals are enforced effectively. His advocacy work has led to stricter regulations against illegal pet breeding and more stringent measures to prevent animal cruelty.

One of his major collaborations has been with a municipal corporation to set up vaccination booths for stray animals. This initiative has significantly reduced the number of rabies cases in the community. He has also worked with NGOs to conduct adoption drives, helping abandoned and rescued animals find permanent homes.

Through his legal advocacy efforts, Dr. Chauhan has contributed to pushing for better infrastructure for government-run animal shelters. He believes that the government must play a more active role in ensuring the well- being of animals, and his relentless lobbying has led to increased funding for shelter facilities.

Disaster Relief and Emergency Response for Animals

Natural disasters often leave animals stranded, injured, and without food. Recognizing this, Dr. Chauhan has been at the forefront of disaster relief efforts, rescuing animals affected by floods, earthquakes, and other calamities. Equipped with a team of volunteers, he ventures into disaster-hit areas, setting up emergency treatment camps and distributing food supplies for animals.

During a major flood in a neighboring district, Dr. Chauhan coordinated with local authorities to rescue stranded cattle, dogs, and birds. His team worked day and night, providing medical aid and ensuring that the animals were safely relocated to higher ground. Such efforts have earned him immense respect from both the animal welfare community and the general public.

Fostering a Culture of Empathy and Volunteerism

Beyond his direct contributions, Dr. Chauhan has played a pivotal role in inspiring young people to take up animal welfare as a cause. He has mentored numerous veterinary students and volunteers, encouraging them to dedicate their skills and time to helping animals. Many of his former students have gone on to establish their own rescue organizations, continuing the legacy of compassion that he has nurtured. His efforts to foster a culture of empathy extend beyond just veterinary students. He has established volunteer programs where ordinary citizens can contribute to animal welfare activities. Whether it's feeding stray animals, assisting in medical camps, or helping with adoptions, Dr. Chauhan ensures that there is a role for everyone who wants to make a difference.

Advocating for Stricter Animal Protection Laws

Recognizing that systemic change is essential for long-term animal welfare, Dr. Chauhan has actively advocated for stronger animal protection laws. He has participated in policy discussions, presented case studies to lawmakers, and pushed for harsher penalties against those found guilty of animal abuse.

One of his biggest legal victories came when he successfully campaigned for a ban on the use of certain harmful drugs in livestock that negatively impact vultures, leading to their declining population. By highlighting the ecological importance of vultures in waste management, he convinced authorities to take action, demonstrating how veterinary science and conservation go hand in hand.

A Legacy of Compassion and Service

Dr. Chauhan's community engagements in animal welfare have not only saved countless animal lives but have also changed the way society perceives and treats animals. His work is a testament to the power of dedication and compassion. Whether through medical treatment, rescue operations, public education, or legal advocacy, he continues to be a guiding force in the fight for animal rights.

His story serves as an inspiration for those who believe that even small actions can create a significant impact. Through his relentless efforts, Dr. Chauhan has built a community that values and protects its animals, ensuring that they receive the care, respect, and dignity they deserve. As he continues his mission, his unwavering commitment to animal welfare remains a beacon of hope for animals and animal lovers alike.

Dr. Chauhan is a dedicated veterinary doctor known for his extensive charitable work in the field of animal welfare. Over the years, he has committed himself to rescuing, treating, and rehabilitating injured, abandoned, and mistreated animals, making a significant impact on countless lives. His work extends beyond just medical treatment; he is deeply involved in spreading awareness about responsible pet ownership, animal rights, and the need for compassion towards all living beings.

Dr. Chauhan's journey into animal welfare began early in his career when he realized that many animals, particularly strays and working animals, were left to suffer without proper medical attention. Determined to make a difference, he started offering free treatment to animals in need, often going out of his way to reach remote areas where veterinary services were unavailable. His tireless efforts soon led him to establish a network of volunteers and animal lovers who assist in identifying and rescuing distressed animals.

One of Dr. Chauhan's most notable contributions is his work with stray animals. He has personally rescued and treated hundreds of dogs and cats, ensuring they receive vaccinations, sterilization, and, when possible, loving homes. His emphasis on sterilization programs has helped control the stray population humanely, reducing instances of abandonment and cruelty.

In addition to companion animals, Dr. Chauhan has extended his expertise to the welfare of farm and working animals. He has treated injured cattle, horses, and donkeys, many of whom were abandoned by their owners after sustaining injuries. His mobile veterinary camps, organized in rural areas, provide much-needed medical care to animals that would otherwise have no access to a veterinarian. Farmers and animal owners are also educated on proper care, nutrition, and humane treatment practices, ensuring long-term benefits for both the animals and their communities.

Despite facing numerous challenges, including a lack of resources, resistance from certain communities, and the emotional toll of dealing with severe cases of neglect and abuse, Dr. Chauhan remains undeterred. He believes that true change can only come through education and consistent efforts, and he has dedicated himself to training young veterinarians and volunteers who share his passion for animal welfare. Through workshops and mentorship programs, he instills in them the values of compassion, dedication, and ethical veterinary practice.

Dr. Chauhan's work has not gone unnoticed. Over the years, he has received recognition from various animal welfare organizations, veterinary associations, and local communities for his selfless service. However, for him, the greatest reward remains the sight of a once- suffering animal regaining health and happiness.

Even as he continues his mission, Dr. Chauhan remains focused on expanding his reach. His vision is to establish a well-equipped animal hospital and shelter where rescued animals can receive long-term care and rehabilitation. He also aims to collaborate with policymakers to strengthen animal protection laws and ensure better enforcement against cruelty.

Dr. Chauhan's story is one of unwavering dedication and deep empathy. Through his work, he has not only saved lives but has also inspired many others to step up for those who cannot speak for themselves. His journey serves as a powerful reminder that a single individual's efforts, driven by kindness and determination, can bring about meaningful change in the world.

On Cowpathy

Dr. Chauhan, affectionately known as the "Cowpathy Man," has become a symbol of compassion and innovative veterinary care in India. With over three decades dedicated to bovine health, Dr. Chauhan has transformed the lives of countless cows and garnered respect from farmers and animal lovers alike. His pioneering work in Cowpathy—a holistic approach to treating cows using natural remedies derived from cow products—has revolutionized the way veterinary medicine is perceived and practiced in rural communities.

Born into a farming family, Dr. Chauhan developed a deep bond with cows from an early age. He witnessed firsthand the crucial role; these gentle creatures play in agricultural sustainability and rural livelihoods. Determined to improve their health and welfare, he pursued veterinary science, graduating with top honors from a reputed institution. However, his journey was just beginning.

In his early career, Dr. Chauhan noticed that conventional veterinary treatments, while effective, often came with side effects and were financially burdensome for impoverished farmers. Driven by the desire to provide accessible and affordable care, he delved into traditional knowledge and Ayurvedic texts, where he discovered the ancient principles of Cowpathy. The method advocates using panchgavya—products obtained from cows, such as milk, urine, dung, ghee, and curd—to treat various ailments.

Dr. Chauhan's research led to groundbreaking treatments for common human diseases, such as lower immunity, cancer and digestive disorders. He formulated herbal concoctions combining cow urine and medicinal herbs that demonstrated remarkable efficacy in boosting immunity and curing infections. His "Gau-Amrit," a tonic made from cow urine, became a popular remedy known for its antibacterial and anti- inflammatory properties.

Beyond medicine, Dr. Chauhan's approach emphasized preventive care. He educated farmers on proper nutrition, hygiene, and shelter for cows, reducing disease incidence significantly. His mobile veterinary camps, which traveled to remote villages, offered free check-ups, vaccinations, and workshops on Cowpathy, making healthcare accessible to thousands of cattle owners.

Dr. Chauhan's holistic methods earned him the nickname "Cowpathy Man," and his reputation spread across states. His contributions were recognized by agricultural universities, and he was invited to deliver lectures, training aspiring veterinarians in Cowpathy. Additionally, his work was featured in various agricultural journals and television programs, inspiring a new generation of veterinarians to adopt sustainable and humane practices. However, Dr. Chauhan's commitment extended beyond healthcare. He played a pivotal role in establishing cow shelters (gaushalas) that served as sanctuaries for abandoned and aging cattle. These shelters not only provided a safe haven but also promoted the use of cow products for organic farming and natural fertilizers, fostering environmental sustainability.

Despite facing skepticism and resistance from conventional veterinary circles, Dr. Chauhan remained steadfast in his mission. His perseverance paid off as empirical evidence and farmer testimonials validated the efficacy of Cowpathy. Today, his treatments are widely accepted, and his model has been integrated into state-run veterinary programs in several regions.

Dr. Chauhan's legacy lies not only in his medical innovations but also in the compassionate ethos he championed. By treating cows with respect and care, he highlighted their indispensable role in rural economies and ecological balance. His story is a testament to the impact one individual can have by blending traditional wisdom with modern science for the greater good.

As the "Cowpathy Man," Dr. Chauhan has redefined veterinary care, ensuring that cows—revered as sacred and vital to agrarian life—receive the dignity and healthcare they deserve. His work continues to inspire efforts towards sust on society.

12

Peer Review: What Others Say About Dr. R.S. Chauhan

COL LEGE OF APPLIED EDUCATION & HEALTH SCIENCES

(Affiliated to Ch. Charan Singh University, Meerut & Recognised by IAP (No. 4154) & UGC, New Delhi) Gangotri, Roorkee Road, Meerut (U.P.)

Dr. S.K. Garg

M.S., Ph.D. (Illinois, USA), FSIIP Director

(Former Vice Chancellor Vety. University. Mathura)

I have known Dr R.S. Chauhan al through his professional carrier. I came in his contact when he was a student, a research worker, a professional colleague. He is very hard working, sincere and dedicated scientist and teacher. I observed his remarkable skills, unwavering dedication and strong commitment for upliftment of research and the profession. He is known as an excellent teacher. He devised simple understandable means for complex topics.

His work on Panchgavya has won lots of appreciation at all levels. In fact he is a person who proved scientifically the application of components of Panchgavya on control of some untreatable disease like cancer.

I have absolutely no hesitation in saying that he is the "Father of Cowpathy". He proved the efficacy and application of products of cow origin with scientific validation.

To be true, even today whenever I have some query, I invariably consult him because I am of the opinion that he has a very strong basic background.

May God always shower his blessings on him.

(S.K. Garg)

Dr. V. Narasinga Rao

M.V.Sc., Ph.D., F.N.A.V.S.
Professor & Head, Vety Medicine (Retd.)
Rajiv Gandhi Institute of Vety. Education
and Research Pondicherry

This is to certify that Dr (Prof) R.S..Chauhan is known to me as an undergraduate student of G.B.pant University of Agriculture and Tech (Pant nagar). He was found to be very hardworking and obedient student with great passion for higher learning. Though I left the University in 1987 on promotion to other universities, our contact continued till now at scientific level. He had done pioneering work on 'Panchagavya' especially their immunomodulatory properties. This detailed work received lot of recognition and appreciation from the scientific world. He has been popular as' Cowpathy Man' in the scientific world. It is worthwhile to note that he acted as Advisor to World Health organization, too He deserves all appreciation and encouragement and I whole heartedly wish him to have many such achievements in future too, even after his retirement on superannuation.

May God Bless him abundantly.

(V. Narasinga Rao)

Dr YPS Malik
Joint Director
ICAR-IVRI Mukteshwar

Prof. R.S. Chauhan is indeed a remarkable figure in the field of veterinary science, particularly known for his pioneering contributions to Cowpathy. His book, "The Panchgavya Cowpathy," delves into his extensive research and dedication, which have significantly enriched the field. Here are some key points about Prof. Chauhan and his work

Expertise and Background: Prof. Chauhan is a distinguished veterinarian and pathologist with expertise in toxicology. He has devoted over four decades to veterinary teaching, research, and extension activities.

Pioneering Work: He is renowned for his work on Panchgavya Cowpathy, which integrates traditional knowledge with modern science. His innovative concepts have brought new insights and advancements to this field.

Recognition: His extensive research and unwavering dedication have earned him the title of "Father of Cowpathy" or "Cowpathy Man" in India.

Book Insights: "The Panchgavya Cowpathy" showcases Prof. Chauhan's scientific acumen and deep commitment to the field. It reflects his efforts to bridge traditional healing practices with modern veterinary science, making it a valuable read for those interested in these intersections.

Prof. Chauhan's work continues to inspire many in the field of veterinary science and traditional healing practices. If you're interested in exploring more about Cowpathy and his contributions, this book is certainly a great resource.

YPS Malik
Joint Director
ICAR-IVRI Mukteshwar

Dr Tilak Singh - Class mate

Dr Chauhan took admission in BVSc & AH in the year 1976 at College of Veterinary Sciences in GBPUAT Pantnagar and fortunately, became his class mate which remained continued till date as friend and residing in same city Rudrapur. Father and elder brother of Dr Chauhan were in Indian Army and I am also fortune to met both of them during our study period. We both came from rural environment and as is evident India lives in villages. We were living in Nehru bhawan hostel's last wing, Dr Chauhan in room no. 46 with Dr AN Singh and Dr YP Singh as his room partners and me in same wing in room no. 52 with Dr YK Sarswat and Dr Devendra Singh Rajput.

Since our student life we were good friends. And we both were considered with high moral and character values which were most of the time compare themselves with the academic toppers Dr NK Verma and Prem Raj Tyagi. We were also attending RSS Shivaji shakha with its very beginning after the emergency which remained continued till date as RSS workers. The RSS shakha gave us moral boost with Sanskar that leads to the honest officers who gave their services to the nation for nation building. Presently Dr Chauhan is known as COWPATHY MAN by virtue of his dedicated research on the subject in spite of all odds.

Dr Chauhan as Scientific Advisor to WHO, Director IVRI, Joint Director CADRAD, Director IBT, Professor and Head Veterinary Pathology motivated several students who were doing wonders in their field in India and abroad. What I should say about Dr Chauhan is loke lighting a lamp in front of sun. Dr Chauhan proved like his name "*yatho namah tatho gun*". And even today he is involved in Gauseva and opened a demonstration unit of Gauma Farm at Rudrapur for the benefit of common man, farmers, animal owners, gaushala workers, veterinarians and medicos besides doing natural farming. The name of Dr Chauhan is in itself is an introduction which is like this saying *"Om ram rameti rameti rame rame manorame, sahasranam tantulyamram nam varanine."*

Views of a class mate!

Dr Tilak Singh
Ex- CVO Moradabad
Department of Animal Husbandry
Government of Uttar Pradesh

Dr Bhanu Singh - Student

It is a proud moment for me to know that Dr R S Chauhan has been recognized as "The Cowpathy Man". I am honored to write about Dr. Chauhan. A short page cannot do justice to his remarkable career. I am fortunate to be one of the many individuals who have benefited from his wisdom and guidance. His passion for uplifting others, combined with his numerous achievements, makes him a true role model, mentor, and teacher.

I first met Dr. Chauhan when he taught the General Veterinary Pathology course at Pantnagar. He inspired me to pursue research in veterinary pathology and instilled in me a passion for science. After completing my veterinary degree and initial research under his supervision, I had the opportunity to collaborate with him on multiple personal and professional levels. Dr. Chauhan's profound impact on my professional growth, along with his unwavering commitment to empowering others, makes him an exemplary teacher and guide. A hallmark of his mentorship is the strong community he has built among his past students. It is no coincidence that many of Dr. Chauhan's former students have risen to prominent leadership positions in professional societies, government organizations, academic institutions, and industries. His mentorship has helped populate the pathology specialty with talented and capable individuals across India and around the globe.

Recognizing the uniqueness of traditional knowledge in cowpathy for disease prevention, Dr. Chauhan's initial research focused on its immunomodulatory aspects. He foresaw the need to establish a scientific basis for cowpathy principles. His expertise in chemicals and immunotoxicity led him to focus on immunomodulation with cow urine. In the early 2000s, this was a challenging and controversial topic, met with skepticism. However, preliminary studies led by Dr Chauhan on immunomodulation with cow urine generated significant interest among scientists. The cow urine distillate (Kamdhenu Ark) was found to enhance immunity in mice, increase the phagocytic activity of macrophages, and stimulate the secretion of interleukin 1 and 2. This seminal work (Chauhan RS, Singh BP, and Singhal LK. *Immunomodulation with Kamdhenu Ark in Mice. Journal of Immunology and Immunopathology,* 3: 74-77, 2001) has been referenced over 100 times by researchers, signifying its scientific relevance and influence. I consider myself fortunate to have worked closely with him on this early research project.

With his vast experience and unwavering commitment, Dr. Chauhan has laid the very foundation of cow therapy in India. Anyone who has interacted with him knows his passion and dedication to this field. His exceptional mentorship has nurtured a new generation of researchers in India. His leadership has resulted in multiple scientific publications, books, and presentations on cow therapy at various platforms. His published research has provided direction for the future of cow therapy. His legacy will continue to shape the field for years to come. It is a proud and well-deserved moment that such a legend is being recognized as the "The Cowpathy Man". This honor is a testament to his extraordinary contributions and his efforts to advance scientific research in the field of cow therapy.

Training under Dr. Chauhan has been one of the highlights of my career, and I take great pride in it. For all that he has taught me and for the immense impact he has had on my life, I will be forever grateful. He is an inspiration, a guiding force, and, of course, a true Guru!

गुरूर ब्रह्मा, गुरूर विष्णु, गुरूर देवो महेश्वरः

Dr. Bhanu Pratap Singh

BVSc, MS, Diplomate ACVP, Diplomate ABT, Fellow IATP
Senior Director, Nonclinical Safety and Pathobiology
Gilead Sciences, Inc.
333 Lakeside Drive, Foster City, CA, USA

Dr Ravindra PV - Student

Dean

It about my mentor and guide, Dr. Ramswaroop Singh Chauhan. To me, Dr. Chauhan is best described as "Ram ke Seva Karnewale, Ram ke Krupa se Ram ke message ko scientifically logon tak pahunchaanewale Vaignanik evam Professor"—a dedicated scientist who translates spiritual wisdom into scientific understanding. Dr. Chauhan is an exemplary scientist characterized by his clarity of thought, decisiveness, honesty, adaptability, and technological savvy. Always willing to learn and share knowledge, he combines profound care for his students with a leadership style defined by fairness, encouragement, and mentorship. His unique personality harmoniously integrates evidence-based scientific reasoning, philosophical insight, and traditional knowledge. To the best of my knowledge, Dr. Chauhan is the pioneering scientist who highlighted the immunomodulatory properties of indigenous (Desi) cow urine and introduced the term **"Cowpathy"** in India. His groundbreaking research provided the first scientific evidence comparing the metabolites in urine from indigenous cows with exotic breeds, establishing its immunomodulatory effects and its potential therapeutic value. He is also credited with introducing the term **"No Observable Effect Level" (NOEL)** dose to scientifically measure toxicological effects of pesticides at doses previously considered safe, revealing hidden impacts on immune cells.

I first met Dr. Chauhan when I joined GB Pant University of Agriculture and Technology, Pantnagar, for my Master's degree (MVSc.). Before joining his team, I was anxious and unsure about research methodologies—how to formulate research questions, perform experiments, analyze data, and interpret results. Under his mentorship on the project "Immunopathological Responses in Poultry Induced by Pesticide Residues," I learned these skills systematically. His detailed guidance and unwavering support transformed my academic anxiety into confidence and maturity.

Dr. Chauhan was well-known among students for his high standards and rigorous expectations, and I was fortunate to be among those chosen to work under his guidance. His appreciation for my sincerity and hard work motivated me immensely. I vividly remember presenting my first DNA laddering and annexin V staining results to him, and the profound sense of accomplishment and pride he shared with me.

As Joint Director at the Centre for Animal Disease Research and Diagnosis (CADRAD) at IVRI, Dr. Chauhan continually supported my PhD research ideas and implementation, always keeping his laboratory and office accessible.

Our interactions extended beyond professional discussions, often continuing at his home, strengthening our bond even further. An astute observer, prolific writer, and passionate communicator, Dr. Chauhan inspired me to present my research in Hindi, a language which is not my native, fostering in me a deep pride in my learning of national language. His teachings extended beyond academia, enlightening me about the scientific basis of traditional practices—for instance, understanding the act of "Arati" as a method that integrates mind, body and spirit and purify the environment when welcoming guests. Our mentor-mentee relationship, cultivated during my Master's and PhD journey, has grown into a lifelong connection. Dr. Chauhan remains an ever-present guide and advisor in my life.

I pray that the Almighty grants him and his family health, happiness, and a long, fulfilling life ahead.

Sincerely

(Ravindra)
Dean of Veterinary Medicine
University School of Veterinary Sciences
ARUBA, USA

Dr Desh Deepak Singh - Student
Professor and Head
Veterinary Pathology
DUVASU Mathura

My teacher and my guide: Prof. (Dr.) R.S. Chauhan

Teachers are the pillars of knowledge and wisdom. They not only teach academic lessons but also impart valuable life lessons. Among all the wonderful teachers I have had, my favourite teacher and my guide Prof. (Dr.) R. S. Chauhan, who taught me Veterinary Pathology during my undergraduate classes (B.V.Sc.&A.H.) and guided me for teaching and research in Veterinary Pathology during Master's (M.V.Sc.) at Govind Ballabh Pant University of Agriculture & Technology, Pantnagar, U.S. Nagar, Uttarakhand. The way of teaching, research, disease diagnosis and his caring attitude make him stand out as an exceptional educator.

Prof. Chauhan is a highly experienced teacher, researcher, diagnostician and administrator with a passion for animal disease diagnosis and research. He has been in teaching, research, extension and administration for more than 40 years and holds Doctor of Philosophy (Ph.D.) in Veterinary Sciences with specialization in Veterinary Pathology from G.B. Pant University of Agric. & Tech., Pantnagar. His unique teaching style, writing skill and research attitude with administrative caliber made him favourite teacher. He made even the most complex Veterinary Pathology concepts easy to understand. Whether it's General Veterinary Pathology, Systemic Veterinary Pathology, Special Veterinary Pathology, Avian Pathology, Necropsy, Oncology or Immunopathology, keeping in view the need of study of Veterinary Pathology to become a good Veterinary doctor he taught everything into simple steps, making it friendly for every Veterinary student. His patience and dedication to ensuring that all students understand the lesson were truly remarkable.

Prof. Chauhan, UG and PG classroom was always filled with energy. He encouraged interactive learning and makes sure every student feel heard. He was never quick to judge or scold; instead, he patiently answers questions, no matter how simple or complicated. His ability to keep the students engaged and interested in the subject was something I admire. What sets Prof. Chauhan apart from other professor's is his personal attention to each student's needs. He often stayed after class to provide extra help and was always available for one-on-one discussions. His encouragement and constant support have motivated me to excel in the field of Veterinary Pathology.

Prof. Chauhan is more than just a teacher to me; he is a mentor and a guide. His impact on my life has been profound, and I will always be grateful for the knowledge and wisdom he has shared with me. I am very fortunate to have Prof. Chauhan as my teacher and guide.

D D Singh
Professor and Head, Veterinary Pathology
DUVASU Mathura

Dr Virendra Garg - Student

Dr. Chauhan: A Mentor, Guide, Inspiration, Beacon in Veterinary Pathology and Cow Protection

It is with great respect and admiration that I write this memoir to honor Dr. R.S. Chauhan, a distinguished professor in pathology who left an indelible mark on the field of veterinary pathology. Dr. Chauhan's career spanned decades, during which he became a guiding light to countless students, scholars, and professionals in the veterinary sciences. His contributions not only advanced the understanding of veterinary pathology but also bridged the gap between science and the preservation of cultural heritage, particularly through his groundbreaking work in cow protection.

Dr. Chauhan is more than just an academic; he is a mentor, a guide, and a visionary in the truest sense of the word. His passion for veterinary pathology was evident in every lecture, every research paper, and every interaction with his students. He possesses the rare ability to make complex pathological concepts accessible and engaging, sparking a deep interest in the subject among all who studied under him. As a result, many of his students have gone on to excel in prestigious institutions, organizations, and research facilities, contributing to the global understanding of veterinary medicine.

I first met Dr. Chahan as a Master's student at G.BP.U.A&T , Pantnagar. At that time, I was driven, but still uncertain of my direction. Pathology, with its detailed and often challenging concepts, seemed daunting. Dr. Chauhan, however, had a unique way of transforming these challenges into opportunities for learning. His approach was not just academic; it was personal. He would patiently walk me through difficult concepts, explaining them in ways that made them not only comprehensible but also deeply fascinating. But Dr. Chahan's influence wasn't confined to the classroom. He saw potential in me that I didn't even recognize in myself. It was during one of our many discussions about the future of veterinary medicine that he first suggested I explore the world of immunology. He spoke about the intersection of pathology and immunology with such enthusiasm that it ignited a spark within me. "Immunology is the key to understanding so much more," he said, his eyes lighting up. "It's the future of medicine, both for animals and humans." His words stayed with me, resonating in my mind long after our conversation ended.

One of the most defining moments of my career came when Dr. Chauhan encouraged me to apply for higher studies at renowned American University. I had never seriously considered it before, thinking it was beyond my reach. His words of affirmation were both reassuring and motivating. With his support and guidance, I felt empowered to take the leap and pursue a path that would shape my future. Looking back, I realize that Dr. Chauhan's mentorship was about more than just passing exams or earning degrees. It was about cultivating curiosity, fostering critical thinking, and, above all, inspiring a sense of purpose. His influence led me to the decision to pursue higher studies.

In the journey of life, there are a few people who leave an indelible mark on your path. Dr. Chauhan, is one such person for me. His influence was profound and far-reaching, extending beyond the realms of veterinary pathology and into the personal sphere of my career aspirations. He not only mentored me through the intricacies of pathology but also sparked in me the passion to pursue something even greater — a PhD in immunology. During the time of my PhD journey in immunology, at Purdue University, West Lafayette, USA., I though of Dr. Chauhan often. His words continue to guide me, reminding me that no goal is too big if you have the right mentor by your side. I am eternally grateful for his unwavering support and for helping me see the bigger picture. Without him, I may never have taken the leap into immunology — and for that, I will always be thankful.

Dr. Chauhan's impact went far beyond the academic realm. He played a significant role in advocating for cow protection, a cause that is close to the hearts of many in India and beyond. His work on cow urine, for instance, was groundbreaking. He was one of the forerunner to scientifically describe its potential benefits, demonstrating its use in various traditional and modern medical contexts. Through rigorous research and scientific inquiry, he was able to validate and promote the therapeutic properties of cow urine, bringing it into the light as a valuable resource for health and healing. Dr. Chauhan's commitment to cow protection was not limited to his research; it extended to his daily life and his efforts to educate others on the importance of preserving and respecting cows as vital cultural and ecological assets. His work in this area was not just academic; it was deeply rooted in his belief in the symbiotic relationship between humans and animals, particularly cows, in Indian society.

As a mentor, Dr. Chauhan is a guiding force for many of us. He instilled in his students not only the rigor of scientific inquiry but also the importance of compassion, ethics, and responsibility in their work. He taught us that our work in veterinary medicine was not just about diagnosing diseases or treating animals; it was about understanding the interconnectedness of all living beings and serving as stewards of their health and well-being. Dr. Chauhan has dedicated his life to the betterment of veterinary pathology, the protection of cows, and the training of countless students who continue to carry forward his legacy. His work will continue to inspire generations of veterinarians, scientists, and researchers who seek to make meaningful contributions to both the scientific community and society at large.

Dr. R. S. Chauhan's influence is not just written in the research papers he published or the students he trained, but in the lives he touched, the ideas he nurtured, and the lives he helped heal through his unwavering dedication to the betterment of animal welfare and science. We are forever grateful for his contribution.

Dr. Virendra Kumar Garg
B.V. Sc. (DVM), M.V.Sc., Ph.D. (Purdue)
Small Animal Veterinarian and Business Owner
Animal Hospital in Fairfield. Fairfield, California, USA

13

Cowpathy Publications

1. Ambwani S, Ambwani Tanuj Kumar and **Chauhan RS** (2014) Counteracting effects of cow urine on allethrin induced immunotoxicity and oxidative stress in chicken lymphocytes culture system. *Journal of Immunology and Immunopathology.* **16**: 58-63.
2. Ambwani S, Ambwani TK and **Chauhan RS** (2018) Ameliorating effects of Badri cow urine on cypermethrin induced immunotoxicity and oxidative stress in chicken lymphocyte culture system. *Biosciences Biotechnology Research Asia.* **15** (3): 1-12.
3. Banga RK, Kumar P, Singhal LK, Sharma A and **Chauhan RS**. 2005. Red hill cattle is characters and body measurements. *The Indian Cow*, **3**: 10-14.
4. Banga RK, Singhal LK and **Chauhan RS**. 2005. Cow urine and immunomodulation: An update on cowpathy. *International Journal of Cow Science*, **1**(2): 26-29.
5. Chauhan Divya and **Chauhan R.S**, 2021. Badri Cattle - A Boon to Uttarakhand. In: Cowpathy and Human Health (Eds : Sethi RS, Kaur Gurvinder, Malik Yashpal, Chauhan RS). SIIP, Ludhiana. P_17
6. Chauhan Divya and **Chauhan R.S.** (2022). Cow Science Experts in India - 1. Shree Krishna A well known Cow Lover. *International Journal of Cow Science.* **6**(1):2
7. **Chauhan R S** and Tulsa Devi. 2020. Immunity, Nutrients and Immunomodulation. *International Journal of Emerging Technologies and Innovative Research*, **7** (5):1030-1038.
8. **Chauhan R.S** and Chauhan Divya (2022).Role of macrophages in prevention and control of diseases and their upregulation through Cowpathy. *International Journal of Cow Science.* **6**(1):10- 29.
9. **Chauhan RS** and Joshi A.2022. Cowpathy in cancer management-therapeutics, prevention and control. In: Cowpathy and human health eds, Chauhan RS and Malik YPS) Innovative Publications New Delhi pp 81-110.

10 **Chauhan RS** and Banga RK. 2005. Veterinarians of India-3. Nakul: 4th Pandav and an expert of equines. *The Indian Cow*, **3**: 19.

11. **Chauhan RS** and Dhama K. 2008. Cowpathy in induction of cytokines and its importance in control of human and animal diseases. *In:* Training-cum-Workshop on "Cytokine Assay". (Eds. BC Pal, AK Srivastava, SK Garg, RC Sharma, S Yadav, A Goel, SN Shukla and R Goel). DUVASU, Mathura. pp 117-123.

12. **Chauhan RS** and Janoti Deepak Singh 2012. Immunomodulatory Molecules. In: Biotechnology-Issues, Opportunities and Challenges (Joshi Ankita, Rawat Anita and **Chauhan RS** (Eds), SIIP Patwadangar, pp 51-69.

13. **Chauhan RS** and Pandey AB. 2013. Parvtiya govansh (Badri cow) evam uske utpadon ki mahatta, In: Parvtiya kshetron me adhunik Gopalan (Eds: AB Pandey, SB Shudhakar and BN Sahu), IVRI Izatnagar.pp28-38.

14. **Chauhan RS** and Rana JMS. 2012. Indigenous cattle genetic resource and their by products for sustainable agriculture,animal and human health. In: Biotechnology-Issues, Opportunities and Challenges (Joshi Ankita, Rawat Anita and **Chauhan RS** (Eds), SIIP Patwadangar,pp208-222.

15. **Chauhan RS** and Shahi BN. 2020. Hematology, serum biochemistry and urinalysis of adult Badri cattle. *International Journal of Livestock Research*, **10** (10): 86-91. IF:2.01.

16. **Chauhan RS** and Shahi BN. 2021. Dry dairy (Shushk godhan gaushala). AWBI Ballabhgarh, pp 104.

17. **Chauhan RS** and Singh GK. 2001. Immunomodulation: An overview. *Journal of Immunology and Immunopathology*, **3**(2): 1-15.

18. **Chauhan RS** and Singh GK. 2013. Immunopathology of A1 beta casein cow milk in man and animals. *Journal of Immunology and Immunopathology*, **15**:201-209.

19. **Chauhan RS** and Singhal LK 2006. Harmful effects of pesticides and their control through cowpathy. *International Journal of Cow Science*, **2**(1): 61-70.

20. **Chauhan RS** and Singhal LK. 2004. Palkapya: A legendry in elephant medicine. *The Indian Cow*, **2**: 46-49.

21. **Chauhan RS** and Tulsa Devi.2020. Impact of pesticide residues on animal and human health and its reversal using cowpathy. *The Journal of India Thoutht and Policy Research*, **10** (1):29-51.

22. **Chauhan RS**, Singh BP and Singhal LK. 2001. Immunomodulation with Kamdhenu ark in mice. *Journal of Immunology and Immunopathology,* **3**: 74-77.

23. **Chauhan RS**, Singh DD, Singhal LK and Kumar R. 2004. Effect of cow urine on Interleukin-1 and 2. *Journal of Immunology and Immunopathology,* **6**(S-1): 38-39.

24. **Chauhan RS**, Tulsa Devi, Joshi A and K Priya.2022. Cowpathy in ancient wisdom, peoples belief, evidenced relief and science with precautionary mearures. In: Cowpathy and human health eds, Chauhan RS and Malik YPS) Innovative Publications New Delhi pp 221-226.

25. **Chauhan RS**. 2002. Panchgavya: A wonder medicine from indigenous cow. *Pashudhan,* **17**: 5-7.

26. **Chauhan RS**. 2004. Panchgavya Therapy (Cowpathy): Current status and future directions. *Indian Cow*, **1**: 3-7.

27. **Chauhan RS**. 2005. Cowpathy (Panchgavya Therapy) and immunomodulation. *In:* Use of alternative systems of medicine (Ayurvedic and Homeopathic) in veterinary practice. (Eds. VM Bhuktar, SS Rautmare, RG Bambal and AM Kulkarni). Nagpur. pp 54-57.

28. **Chauhan RS**. 2005. Cowpathy: A new version of ancient science. *Employment News*, **XXX**(15): 1&2.

29. **Chauhan RS**. 2006. Cowpathy induced immunomodulation. *In:* Immunology in Health and Disease. (Eds. MK Agarwal and SK Awasthi). CSJM University, Kanpur. pp 176-179.

30. **Chauhan RS**. 2007. Immunomodulation using cowpathy. *In:* IV SIIP Convention and National Symposium on "Immunobiotechnology" at Chennai on Feb. 25-27, 2007. pp 75-80.

31. **Chauhan RS**. 2007. Nutrition and immunity. *In:* Nutritional and Biotechnological Tools for Economic Livestock Production. (Eds. N Agarwal, LC Chaudhary, AK Verma, DN Kamra and K Sharma). IVRI, Izatnagar. pp 35-40.

32. **Chauhan RS**. 2007. Significance of Panchgavya for animal and human health care: Immunomodulation using cow urine. *In:* Vande Gomatharam National Conference on "Glory of Gomatha" at Tirumala Tirupati Devasthanams, Tirupati (AP) on December 1-3, 2007. pp 76-80.

33. **Chauhan RS**. 2008. Cow urine induced immunomodulation. *In:* Souvenir on "Research on Panca Gavyas". (Eds. Shirley Telles, Naveen KV and KP Ramesh). Govigyan Anusandhan Kendra, Nagpur. pp 55-

56.

34. **Chauhan RS**. 2010. Cowpathy in Cancer Management. *In:* Recent Advances in "Immunobiotechnology". (Ed. RS Chauhan and JMS Rana). IBT, Patwadangar (UK). pp 75-85.
35. **Chauhan RS**. 2013. Indigenous cow urine and Immunomodulation. *Journal of Immunology & Immunopathology*. **15:**19-22.
36. **Chauhan RS**. 2014. Cowpathy and Immunomodulation. *In* National Seminar on "Panchgavya Chikitsa" at CCRAS New Delhi, on August 28-29, 2014.
37. **Chauhan RS**. 2017. Cowpathy and Rural Development. *In* National Seminar on "Continuity of Indian Identity" at Delhi University, on January14-15, 2017.
38. **Chauhan RS**. 2017. Cowpathy and Rural Development. *In:* International Seminar on "Go navratri Mahotsav" at Jagannath Puri on October 26-28, 2017.
39. **Chauhan RS**. 2017. Health effects of A1 andA2 Milk. *In:* International Seminar on "Go navratri Mahotsav" at Jagannath Puri on October 26-28, 2017.
40. **Chauhan RS**. 2018. Current Scenario of Chemical Farming Road Map of Rural Development- The Indian Cow Model. *In:* MANAGE Sponsored Training programme on Capacity Building of Extension Functionaries for Promotion of Entrepreneurship among Farmers, on January1, 2018, DUVASU Mathura.
41. **Chauhan RS.** 2025 Zootherapy and Panchgavya (Cowpathy). In: Veterinary Ayurveda and Traditional Practices, ed: KML Pathak, pp170-179.
42. **Chauhan RS**. 2025. Dry dairy to increase the income of gaushalas and gaupalaks. Animal Welfare Board of India. Annual function on award ceremony, 27 Feb2025 at New Delhi.
43. **Chauhan RS**. And Malik YPS 2022. Cowpathy and Human Health. Innovative publications New Delhi, Pp356.
44. **Chauhan RS.**1999. Manav ko Rog Sonpate Pashu.Kapish Prakashan, Gurgaon.
45. **Chauhan RS.**2004. Govansh evm Manav Shwashthya. Go chitsha samiti ,Pantnagar
46. **Chauhan RS**.2017. Cowpathy : A new version of ancient science. *ICAR NEWS*, **23** (4): 4-5.

47. **Chauhan RS**.2018. Cancer : Therapeutics, prevention and control. *Journal of Immunology & Immunopathology*.**20**: 94-106.

48. **Chauhan RS.**2022. Cow urine and Immunomodulation. In: Cowpathy and human health eds, Chauhan RS and Malik YPS) Innovative Publications New Delhi pp 69-80.

49. **Chauhan RS**.2022. Cowpathy-An Introduction. In: Cowpathy and human health eds, Chauhan RS and Malik YPS) Innovative Publications New Delhi pp 23-36.

50. **Chauhan RS.**2022. Rsearch protocols to study immunomodulatory and anti cancer properties of cow urine distillate in laboratory animals. In: Cowpathy and human health eds, Chauhan RS and Malik YPS) Innovative Publications New Delhi pp 111-140.

51. **Chauhan, R.S.** and Prasad, RB. (2004). Hill cattle and their importance in Uttaranchal. National Symposium on *"Conservation and Propagation of Indigenous Breeds of Cattle & Buffaloes"* held at Pantnagar on Feb 26-28, 2004. pp. 24-26.

52. **Chauhan, R.S.**, Tulsa Devi, Divya Chauhan and Shuchi Verma. 2022. Cow pathy in management of immunity and cancer in animals and man. *International Journal of Current Research*.**14** : 22719-22723

53. **Chauhan, RS** . 2020. Research on Panchgavya- Current status and future directions. In: National Seminar on Panchgavya for Health Care – A Potetial area for Pharmaceutical Research, held on 8-9 Feb 2020 at Anurag College of Pharmacy, Nagpur.

54. **Chauhan, RS** (2023) Cowpathy for one health. In: Progressive Horticulture conference Feb 3-5, 2023. At GBPUAT Pantnagar.

55. **Chauhan, RS**. 2022. Bhartiya gay evm panchgavya gopathy (Ek tatthyatmak vegyanik vishleshan). Go Mahima Seva Trust, Inderpur Rudapur, Uttarakhand.

56. **Chauhan, RS**. 2023. Panchgavya – Vegyanik tathya. Go Mahima Seva Trust, Inderpur Rudapur, Uttarakhand.

57. **Chauhan, RS**.2020. Indigenous cow urine upregulates the immunity. *epashupalan*,May1, 2020.

58. **Chauhan, RS**.2020. Road map of Rural Development of Uttarakhand-The Badri cow model. *epashupalan*,May1, 2020.

59. Dhama K, **Chauhan RS** and Singhal Lokesh. 2005. Anti-cancer activity of cow urine: Current status and future directions. *International Journal of Cow Science,* **1**(2): 1-25.

60. Dhama K, Khurana SK, Karthik K, Tiwari R, Malik YPS and **Chauhan RS**. 2014. Panchgvya: Immune-enhancing and Therapeutic Perspectives. *Journal of Immunology and Immunopathology,* 16: 1-11.

61. Dhama K, Rathore R, **Chauhan RS** and Tomar S. 2005. Panchgavya (Cowpathy): An overview. *International Journal of Cow Science,* **1**(1): 1-15.

62. Dhama K, Rathore R, **Chauhan RS** and Tomar S. 2008. Panchgavya (Cowpathy): An Overview. *The Indian Cow*, **17**: 45-68.

63. Garg N and **Chauhan RS**. 2004. Draught power: Backbone of Indian economy. *The Indian Cow*, **2**: 11-22.

64. Garg N, Kumar A and **Chauhan RS**. 2004. Assessing the effect of cow urine on biochemical parameters of white leghorn layers. *The Indian Cow*, **2**: 38-41.

65. Garg N, Kumar A and **Chauhan RS**. 2005. Effect of Indigenous cow urine on nutrient utilization of white leghorn layers. *International Journal of Cow Science,* **1**(1): 36-38.

66. Garg N, Kumar A, **Chauhan RS**, Singhal LK and Lohni, M. 2004. Effect of cow urine on the production and quality traits of eggs in layers. *The Indian Cow*, **1**: 12-15.

67. Jakhar J and **Chauhan RS.**2022. Natural killer (NK) cells, their functions and upregulation due to cowpathy. In: Cowpathy and human health eds, Chauhan RS and Malik YPS) Innovative Publications New Delhi pp 205-220.

68. Joshi A and **Chauhan RS** (2012). Anticancer effect of Taxus Baccata and Cow Urine on Biochemical attributes of mice treated with Diethyl Nitrosamine. *Indian Journal of veterinary pathology*. 36(2):203-208.

69. Joshi A and **Chauhan RS** (2014). Anti cancer effect of *Taxus baccata* and Indian cow urine distillate (CUD) on mice treated with diethylnitrosamine- Pathmorphological stuty. *International Journal of Scientific and Engineering Research*. **5** (11): 140-144.

70. Joshi A and **Chauhan RS** 2013. Evaluation of Anticancer properties of Taxus baccata and Badri cow urine in mice: Clinicoheamatological study. *International Journal of Advance research*. **1** (5):71-78.

71. Joshi A and **Chauhan RS**. 2013. Anticancer effect of *Taxus baccata* and Badri cow urine on biochemical attributes of mice treated with diethyl nitrosamine. *In:* Journal of Immunology and Immunopathology, Jan-June, 2013 at RAJUVAS Bikaner on December 22-24, 2013. pp 129.

72. Joshi A and **Chauhan RS**. 2013. Evaluation of anticancer properties of *Taxus baccata* and Badri cow urine in mice: Clinico-hematological stuty. *In:* Journal of Immunology and Immunopathology, Jan-June, 2013 at RAJUVAS Bikaner on December 22-24, 2013. pp 129.

73. Joshi A**,** Bankoti K, Bisht T and **Chauhan RS** (2012)**.** Immunomodulatory effect of Gir cow urine distillate in Rabbits. *Journal of Immunology and Immunopathology.* **14**(1): 57-61.

74. Kumar P, Banga RK and **Chauhan RS**. 2005. Red Sindhi: Important dairy breed of the Indian subcontinent. *The Indian Cow*, **3**: 40-42.

75. Kumar P, Singh DD, Singh GK and **Chauhan RS**. 2004. Milch animal wealth of India: An Apraisal. *The Indian Cow*, **1**: 31-34.

76. Kumar P, Singh GK, **Chauhan RS** and Singh DD. 2004. Effect of cow urine on lymphocyte proliferation in developing stage of chicks. *The Indian Cow*, **2**: 3-5.

77. Kumar R, **Chauhan RS**, Singhal LK, Singh, AK and Singh DD. 2002. A comparative study on immunostimulatory effects of Kamdhenu ark and Vasant Kusumakar in mice. *Journal of Immunology and Immunopathology,* **4**(1&2): 104-106.

78. Nandi S, Pandey AB, Manohar M and **Chauhan RS**. 2008. Sero-surveillance of infectious bovine rhinotracheitis in cow bulls and buffalo bulls in India. *Indian J. Comp. Microbiol. Immunol. Infect. Dis.* **28**(1&2): 1-3.

79. Pant Neha and **Chauhan R S**.2020. Pandemic of Infectious Diseases due to New Etiological Agents Predisposing factors, Case study of COVID19 and Control Measures. *International Journal of Current Microbiology and Applied Sciences*, **9** (6) : 3424-3457.

80. Parihar Manisha and **Chauhan RS**. 2020. Health Implications of A1 and A2 Cow Milk Proteins. *International Journal of Creative Research Thoughts*, **8** (5):2383-2397.

81. Raina M Tamrakar V Rathor DPS Joshi A and **Chauhan RS**. 2013. A comparative study of serum biochemical of Sahiwal, Badri, cross bred and Tulsi cows *In* Journal of Immunology and Immunopathology, Jan-June, 2013 at RAJUVAS Bikaner on December 22-24, 2013. pp 104.

82. Rathore A, Singh S, and **Chauhan R.S**. (2023). Effect of Indigenous Badri Cow Urine Distillate of Hill Region on Hematological Profile of Wistar Rats. Poster Presented in Panchagavya Diwas organized by Ramjas college in collaboration with ministry of Fisheries, Animal Husbandary and Dairying. At Ramjas college, University of Delhi on Oct 13,2023.

83. Rathore A, Singh S, and **Chauhan R.S.** (2023). Research on Cow urine validation protocols for determining the immunomodulatory properties. Paper presented in Recent Advances in Panchagavya Research & Innovations, Panchagavya 2023 at Sawangi, DMIHER, Wardha, Maharashtra on March 11-12,2023.

84. Rathore A, Singh S, and **Chauhan R.S**. (2024). Impact of Badri Cow Urine Distillate on Cell-Mediated Immunity of Wistar Rats. Paper presented in One Health Initiative: Harmonizing human, animal & environmental health, (OHI-2023) At GLA University, Mathura on Feb 18-20,2024.

85. Rathore A, Singh S, Devi T, Chauhan D and **Chauhan R.S.** (2023). Bacteriological Quality of Indigenous Badri Cow Urine. *Journal of Immunology and Immunopathology*, **25** (2):113-116.

86. Rathore A, Singh S, Devi T, Chauhan D and **Chauhan R.S**. (2023). Bacteriological Quality of Indigenous Badri Cow Urine. Paper Presented in National conference on Gau – Vigyan in Modern Life and Medical Science (NCGV–2023) at IIT, Guwahati, Assam on May 20 to 21,2023.

87. Rathore A, Singh S, Kamboj A, and **Chauhan R.S.** (2024). Immunomodulatory Property Of A Herbal Preparation Containing Badri Cow Urine with Herbs. Paper presented in National seminar on Indigenous cow, organic farming and panchagavya chikitsa (NSIOP 2024), At BHU, Varanasi on Sep 21-22,2024.

88. Ravindra PV, Khan D, Umapati V and **Chauhan RS**. 2004. New weapons of prevention and control of diseases from myth to reality. *Livestock International*, **8**(6): 17-20.

89. Saxena S, Garg V and **Chauhan RS**. 2004. Cow urine therapy: Promising cure for human ailments. *The Indian Cow*, **1**: 25-30.

90. Sethi RS, Kaur G, Malik YPS and Chauhan,RS.2022. Cowpathy and human health compendium. SIIP Pantnagar – GADVASU Ludhiana.60pp

91. Shahi BN and **Chauhan RS.**2022. Badri cow habitat, breeding and management for quality cowpathy products. In: Cowpathy and human health eds, Chauhan RS and Malik YPS) Innovative Publications New Delhi pp 149-154.

92. Singh AK and **Chauhan RS**. 2004. Haemorrhagic septicemia in cattle: A cause of worry to farmers. *The Indian Cow*, **2**: 23-26.

93. Singh BP and **Chauhan RS**. 2004. Cow dahi (Curd) or Matha (Butter milk): As proiotic to control animal diseases. *The Indian Cow*, **2**: 6-10.

94. Singh BP and **Chauhan RS**. 2008. Cow dahi (Curd) or matha (Butter milk): As probiotic to control animal diseases. *The Indian Cow*, **15**: 42-45.

95. Singh BP, **Chauhan RS** and Singhal LK. 2003. Toll Likc Receptors (TLRs) and their role in innate immunity. *Current Science*, **85**(8): 1156-1164.

96. Singh DD, Kumar P and **Chauhan RS**. 2004. Comparative profiles of indigenous and exotic cows. *The Indian Cow*, **1**: 56-60.

97. Singh DD, Kumar P and **Chauhan RS**. 2004. Veterinary scientists of India-1: Salihotra- Father of Veterinary Science. *The Indian Cow*, **1**: 64-65.

98. Singh DD, Singhal LK, Garg N, Kumar R and **Chauhan RS**. 2003. Kamdhenu Ark enhances humoral immunity in rats. *In:* National Symposium on Molecular Biology in India: A Postgraduate Update held at Gwalior on January 18, 2003.

99. Singh GK and **Chauhan RS**. 2002. Specific and paraspecific immunity in developmental stages. *Pantnagar News,* **5**(2): 3-4.

100. Singh R and **Chauhan RS**. 2008. Hygienic production of bovine semen. *The Indian Cow*, **16**: 57-61.

101. Singh S, Rathore A and **Chauhan R.S**. (2023). Evaluating the impact of Badri cow urine distillate on the serum biochemical profile of Wistar rat. Poster Presented in Panchagavya Diwas organized by Ramjas college in collaboration with ministry of Fisheries, Animal Husbandary and Dairying. At Ramjas college, University of Delhi on Oct 13,2023.

102. Singh S, Rathore A and **Chauhan R.S.** (2024). Evaluating the impact of Badri cow urine distillate on the serum biochemical profile of Wistar rat *Journal of Immunology and Immunopathology*, **26** (1): 70-77.

103. Singh S, Rathore A and **Chauhan R.S.** (2024). Impact of Badri Cow Urine on Humoral Immune Response in Wistar Rats. *Journal of Immunology and Immunopathology*, **26**(1):78-83.

104. Singh S, Rathore A and **Chauhan R.S.** (2024). Impact of Badri Cow Urine on Humoral Immune Response in Wistar Rats. Paper presented in One Health Initiative: Harmonizing human, animal & environmental health, (OHI-2023) At GLA University, Mathura on Feb 18-20,2024.

105. Singh S, Rathore A, Devi T, Chauhan D and **Chauhan R.S.** (2023). Indigenous Badri Bull Urine: A Bacteriological Analysis. Poster Presented in National conference on Gau – Vigyan in Modern Life and

Medical Science (NCGV–2023) at IIT, Guwahati, Assam on May 20 to 21,2023.

106. Singh SK, Bist B and **Chauhan RS**. 2004. Seroprevalence of brucellosis in Gaushala of Mathura district and its public health significance. *Journal of Immunology and Immunopathology,* **6**(S-1): 131-132.

107. Singh SK, **Chauhan RS**, Singh MP and Singh R. 2005. Indigenous disease control practices among livestock owners of Chambal region of Agra district (UP) effect of seasons on osmotic fragility and haematological indices of crossbred cows. *The Indian Cow*, **3**: 32-39.

108. Singh Sriprakash, Kumar P, Singh GK, **Chauhan RS** and Saxena S. 2004. Sahiwal: Most popular cow of the subcontinent. *The Indian Cow*, **1**: 35-38.

109. Singhal L and **Chauhan RS**. 2007. Indigenous cattle breed- Tharparkar. *International Journal of Cow Science,* **2**(2): 71.

110. Singhal LK, Singh B and **Chauhan RS**. 2007. Immunomodulatory effects of cow urine. *The Indian Cow*, **12**: 2-4.

111. Sinha DK, Garg AK and **Chauhan RS. 2007**. Govansh evm Panchgavya. IVRI Izatnagar.

112. Sohal JS, Amit Kumar, Verma AK, Kumar Aruna T, Malik YPS and **Chauhan RS**.2020. Immunology in 21st century for Improvising one health. SIIP Pantnagar/ Meerut.

113. Tamrakar V, Raina Manveen, Rathor DPS, Joshi Ankita and **Chauhan RS** (2015) Comparative analysis of serum biochemical of Sahiwal, Badri, cross-bred and Tulsi cows. *Journal of Immunology and Immunopathology.* **17**: 86-91.

114. Tulsa Devi, Chauhan Divya and **Chauhan R.S** (2022) Immunomodulatory activity of Badri Cow urine distillate and Tulsi (*Ocimum sanctum*) extract on wistar rats. Presented in National seminar on indigenous Cow, Agriculture and holistic health at B.H.U Varanasi on 5 to 7 August 2022.

115. Tulsa Devi, Prachi Jaiswal, BN Shahi and **RS Chauhan** . 2020. Badri cow urine: Chemical finger printing and Immunomodulation. Presented in International e conference on Immunology in 21st century for improvising one- health. August 7-8,2020. (Awarded with best poster paper award)

116. Verma Jiya and **Chauhan RS**. 2020. COVID - 19 Apocalypse: Therapeutic and Preventive Measures for its Containment in India. *International Journal of Life Sciences,* **8** (2) :327-341.

13

Some Important Moments in The Eyes of Camera

Receiving student award from Dr IP Singh Dean Veterinary Sciences

Receiving student award from Mrs (Dr) Shinade Acting Vice-Chancellor

Dr Rajendra Prasad award of ICAR-2002 (Union Agriculture Minister Shri Ajit Singh is giving award to Prof RS Chauhan in the presence of Shri Hukum Dev Narayan Yadav Minister of state and Secretary agriculture)

Receiving "Gopal Gaurav award" from Shankaracharya Shri Raghvendra Saraswati ji maharaj of Shri Ramchandrapuram muth, Shimoga Karnataka

''गोपाल गौरव'' सम्मान प्राप्त करते हुए

Dr Jeff G Vos Congratulating Prof RS Chauhan on becoming advisor WHO at NIPHE Bilthoven, The Netherlands.

Dr RS Chauhan is with the Union Agriculture minister Shri Sompal Shashtri ji in a function of Prakrati Bharti at Kota (Rajasthan)

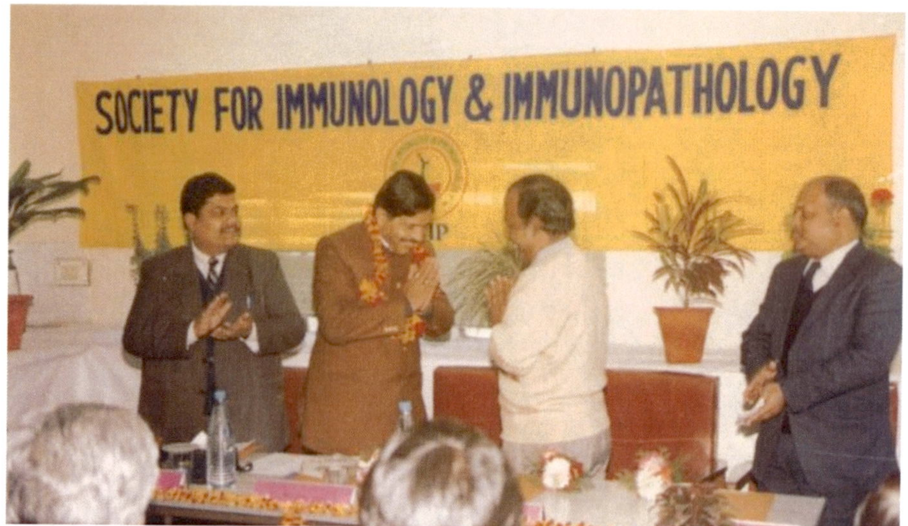

Dr RS Chauhan is with the Union Science & Technology minister Shri Bachi Singh Rawat ji in a function of SIIP at CIMAP Lal kuan (Uttarakhand)

Dr RS Chauhan is with Union Agriculture Minister Shri Rajnath Singh ji at Pusa New Delhi

Dr RS Chauhan is Receiving Gau Seva appreciation award from Union Animal Husbandry Minister Shri Purushottam Rupala ji at New Delhi

A book written/ edited by Dr RS Chauhan and Dr YPS Malik is releasedby Union Animal Husbandry Minister Shri Purushottam Rupala ji at New Delhi

DDG (AS) is with Dr RS Chauhan in his laboratory looking anticancer effect of Badri cow urine in cell culture under microscope.

Dr RS Chauhan is with Dr CM Singh (President VCI), Dr NK Ganguly DG (ICMR) and Dr A. Kumar Head Microbiology AIIMS during inaugural function of SIIP.

Dr HR Khanna Joint Commissioner (AH) honoring Dr RS Chauhan during AWBI function on 27-02-2025 at New Delhi.

Dr RS Chauhan presenting books on Cowpathy to the Minister Dr SPS Baghel during AWBI function on 27-02-2025 at New Delhi.

Dr Triveni Dutt Director ICAR-IVRI is honoring Dr RS Chauhan with Life Time Achievement award of Indian Association of Veterinary Pathologists, along with Dr BN Tripathi DDG(AS), President IAVP and Dr C Balachandran VC TANUVAS during.

Dr AK Srivastav VC DUVASU is honoring Dr RS Chauhan with Life Time Achievement award of Society for Immunology and Immunopathology, along with Dr BN Tripathi VC Jammu, President IAVP ; Dr GS Tomar President SIIP, Dr P Malik AHC and Dr Manish Director ICAR-CIRG.

Dr RS Chauhan is Shri Biswash ji HE Governor of Chhattisgarh

Dr RS Chauhan is Discussing Immunopathological problems with trainees of DBT Sponsored training programme on Immunopathology.

Dr RS Chauhan is with VC HAU Hisar and Km Shailjia ji MOS Education (Govt of India) during inaugural function of II IAAVR conference and release of book "Veterinary Clinical and Laboratory Diagnosis.

Dr RS Chauhan is with Shri Bachi Singh Rawat ji MOS Science & Technology (Govt of India) shaking hands with my father Shri Rajpal Singh during inaugural function of release of First issue of Journal of Immunology and Immunopathology.

Dr RS Chauhan organized a training programme on Dot Immunobinding Assay (DIA)along with the experts

Dr RS Chauhan is receiving Padma Shri Samman in a function at Hotel Ashok New Delhi